T0261220

EDITED BY
PAUL MARTIN, STEVIENNA DE SAILLE,
KIRSTY LIDDIARD AND WARREN PEARCE

BEING HUMAN DURING COVID-19

BRISTOL
UNIVERSITY
PRESS

First published in Great Britain in 2022 by

Bristol University Press
University of Bristol
1-9 Old Park Hill
Bristol
BS2 8BB
UK
t: +44 (0)117 374 6645
e: bup-info@bristol.ac.uk

Details of international sales and distribution partners are available at
bristoluniversitypress.co.uk

British Library Cataloguing in Publication Data
A catalogue record for this book is available from the British Library

ISBN 978-1-5292-2312-5 hardcover
ISBN 978-1-5292-2313-2 ePub
ISBN 978-1-5292-2314-9 ePdf

Cover design: Bristol University Press
Front cover image: iStock-1302892765

Printed and bound by CPI Group (UK) Ltd, Croydon, CR0 4YY

Contents

List of Figures and Table

Figures

Table

Notes on Contributors

Giorgia Aiello is Professor of Culture and Communication at the University of Leeds and Associate Professor in Sociology of Culture and Communication at the University of Bologna.

Effie Amanatidou is a research and innovation policy analyst with a special interest in transnational cooperation, evaluation, and impact assessment of research, social innovation, and foresight.

C.W. Anderson is Professor of Media and Communication at the University of Leeds.

Keren Naa Abeka Arthur is Senior Lecturer and Director of the University of Cape Coast D-Hub. She has a PhD and MBA from the University of Exeter.

Rokia Ballo is an STS PhD student at UCL and Co-Chair of Science London. Her research focuses on the construction of science for policy and social inequalities in the United Kingdom.

Timothy Birabi is a quantitative economist with a strong background in data analytics, financial modelling and corporate strategy. He is a consultant based in London.

Carlos Cuevas-Garcia is a postdoctoral researcher at the Technical University of Munich. His research explores the interrelations between innovation, research, and identity configurations.

Stevienna de Saille is Lecturer in Sociology at the University of Sheffield working in the intersection of science and technology studies, social movement theory, and heterodox economics, with an emphasis on responsible innovation of emerging technology.

Katy Evans works as an associate for Changing Our Lives, a rights-based organisation which champions the rights of disabled people and people with mental health difficulties to live ordinary lives. She is a co-researcher in the Living Life to the Fullest project.

Dan Goodley is Professor of Disability Studies and Education at the University of Sheffield and Co-Director of iHuman. He is a Nottingham Forest fan and Dad to two young women.

Dawn Goodwin is Senior Lecturer at Lancaster University where she teaches sociology of health and illness. Her recent research focuses on public inquiries in healthcare.

Helen Kennedy is Professor of Digital Society in the Department of Sociological Studies, University of Sheffield, where she researches how developments in digital technology are experienced in everyday life.

Kirsty Liddiard is Senior Research Fellow in the School of Education and iHuman at the University of Sheffield. Her chapter is co-authored with the Co-Researcher Collective, a group of expert-by-experience researchers.

Paul Martin is Professor of the Sociology of Science and Technology at the University of Sheffield. He researches the pharmaceutical industry and the development of genomic medicine.

Rod Michalko is retired from the University of Toronto where he taught disability studies. He is the author of numerous articles and five books.

Michael Morrison is a qualitative social scientist, drawing on science and technology studies and medical sociology to explore the social shaping of emerging biomedical technologies.

Brigitte Nerlich is Professor Emeritus of Science, Language, and Society at the University of Nottingham. She studies the role of metaphors in debates about science.

Poonam Pandey is Assistant Professor of Public Policy at FLAME University, Pune. Her work focuses on responsible innovation and democratic governance of emerging technologies in the Global South.

Warren Pearce is Senior Lecturer at iHuman and the Department of Sociological Studies, University of Sheffield. He teaches and researches digital methods and how science is used in public debates about politics and policy.

Anna Pilson is a PhD candidate at the University of Durham School of Education using participatory and autoethnographic methods with visually impaired young people to conduct research.

Paul Graham Raven is Marie Sklodowska-Curie Postdoctoral Fellow at Lund University, Sweden, working with the narrative rhetorics of sociotechnical and climate imaginaries.

Camilla Mørk Røstvik is Lecturer in Modern and Contemporary Art History at the University of Aberdeen.

Katherine Runswick-Cole is Professor of Education in the School of Education at the University of Sheffield. Katherine's work is located in critical disability studies, exposing and challenging the processes of dis/ableism.

Aviram Sharma is Assistant Professor at the School of Ecology and Environment Studies at Nalanda University. His research primarily employs an interdisciplinary approach and lies on the interface of science and technology studies, environmental studies, public understanding of science, and heterodox economics.

Cliff Shelton is Consultant Anaesthetist at Wythenshawe Hospital and Senior Clinical Lecturer at Lancaster Medical School.

Ruth Spurr is a blogger and part-time model for Zebedee. She blogs about illness, youth, young people's lives and her assistance dog, Willow. She is a co-researcher in the Living Life to the Fullest project.

Tanya Titchkosky is Professor of Disability Studies in the Department of Social Justice Education at OISE at the University of Toronto. Her teaching and writing is committed to an interpretive inter-relational approach.

Stefania Vicari is Senior Lecturer in Digital Sociology at the University of Sheffield.

Emma Vogelmann is a leading activist in the disabled community. She is Lead Policy Adviser at Scope, on the children and young people team, where she focuses on the issues faced by disabled children, young people and their families. She is a co-researcher in the Living Life to the Fullest project.

Lucy Watts MBE is a proud young disabled woman with a passion for the great outdoors, for writing and photography, who dedicates her time to making a difference for others and lives life to the fullest with support from her assistance dog Molly. She is a co-researcher in the Living Life to the Fullest project.

Kate Weiner is Senior Lecturer in Sociology at the University of Sheffield. She researches the construction of medical knowledge and everyday health practices.

Sally Whitney is a community researcher with a specialist interest in the lives of disabled young people, their access to work, and the impact of assistance dogs in their lives. She is a co-researcher in the Living Life to the Fullest project.

Ros Williams is a Wellcome Trust research fellow at the University of Sheffield. Her research explores ideas of 'race', biomedicine, and the technologies that connect them.

Zheng Yang is Assistant Professor in the School of Communication, Soochow University. He got his PhD degree in sociology from the University of Sheffield.

Introduction

Paul Martin, Stevienna de Saille, Kirsty Liddiard
and Warren Pearce

This book centres on questions of the human that are raised by the pandemic which began in 2019 and addresses these through a series of short, accessible and thought-provoking essays that range across disciplinary boundaries and intellectual silos.

The COVID-19 crisis poses massive challenges for many citizens, businesses, policymakers and professionals around the globe. The pandemic has highlighted the deep divisions and inequalities that already existed, while at the same time opening up new fissures and fractures in society. However, as many have commented, the crisis also presents new opportunities to fundamentally rethink many aspects of social, cultural, psychological and economic life. Three key issues have emerged in this context that are fundamentally concerned with the experience, meaning and understanding of being human. *Firstly, the marginalization of many groups of people*, most notably members of Black, Asian and minority ethnic (BAME) communities, disabled, young, older and displaced people and how they are de/valued in the response to the virus. It is vital that their experiences are included when thinking about life after COVID-19. This collection pays special attention to the experience of disabled people, a group often neglected in many discussions of the pandemic. *Secondly, the role of new scientific knowledge* and other forms of expertise in these processes of inclusion and exclusion. Little critical attention has so far been paid to the central role of science in shaping our understanding and experience of the pandemic. *Thirdly, the remaking and reordering of society as a result of the pandemic and the opening up of new futures* for work, the environment, culture and daily life. At the same time, the relevance and applicability of human and social sciences have been debated as we enter a period of knowledge generation that has emphasized the biomedical over the socio-political or psycho-political. These critical understandings of how we might better make the future are still missing from public discussion of the post–COVID-19 world.

The focus on 'the human' as a central analytical heuristic is a defining feature of the approach taken here. This owes much to ideas coming out of the broader field of what has become known as the critical posthumanities. It is a diverse and unruly set of concepts and methods for understanding the human that share a common desire to unsettle traditional assumptions about what it means to be human, who is or is not in this category, and decentring the human as the focus of both scholarly attention and political practice. It also emphasizes our entanglements with both other humans and various non-humans, including animals, the environment and technology. However, this collection does not simply just connect with the posthumanities, as it is more eclectic and open to other forms of scholarship grounded in both the social sciences and science, technology, engineering and maths (STEM) disciplines. The benefit of such an approach is in its diversity.

The other focus of the collection is, of course, COVID-19. Here again the starting point for these essays is not simply to chart the experience and consequences of the pandemic but to use COVID-19 as a means of exploring some of the central dynamics of power and knowledge at work in contemporary societies. This sheds light on the way in which the pandemic has reinforced existing inequalities, created new possibilities and remade ideas about the human. The pandemic has clearly transformed society: and we seek to understand this transformation in terms of negation and possibility.

While the authors are from a wide range of backgrounds in the social and human sciences, humanities and STEM disciplines, we share a commitment to a more human future, one that challenges orthodox thinking and seeks to imagine how we can build a better world after the pandemic. The essays draw on empirical studies of policy making and co-produced research drawing on the lived experience of marginalized groups, as well as social media analysis, sociological theory, literature studies and work on new technoscientific knowledge. In doing so they start to address some of the most interesting and important questions about how we understand the contemporary human. These include:

- How is the human being conceived of in different domains by new forms of knowledge, political action and social norms?
- Where in the world is understanding of the human being expanded, contested, pushed back or unmade? Do variations exist in terms of place? Are particular notions of being human specific to the West or Global North?
- How are ideas of the human being regulated by government, private actors and civil society?
- Can we talk about an emerging post–COVID-19 human that is distinct from the pre-pandemic human? What sort of humans do we want to become?

The chapters in this collection do not systematically answer the questions, but they do start to outline how we might address them in a more creative fashion. Orthodox conceptions of the human condition and the category of 'the human' tend to (overtly or tacitly) privilege the normative: white, heterosexual, male, able-bodied and minded, metropolitan, speaking a standard language, located in Western European and North American contexts. Many of the contributors to this book reject this minority world perspective and instead seek to reinsert the perspectives of those living in the majority world who we might describe as being part of the 'Missing Humanities'. These perspectives are largely missing from current debates on the pandemic and its significance and consequences as we think about life after COVID-19. These essays therefore seek to critically intervene to broaden these discussions and envision a more equitable and inclusive human future.

The book is organized under four distinct but related headings, each of which highlights a key aspect of how the pandemic has shaped our understandings of being human and the possibilities this opens up in post-COVID-19 times. This relates to changing representations of the human as a result of new knowledge and media (Knowing Humans); the dynamics of exclusion that have become so visible during the pandemic (Marginalized Humans); new scientific discourses that are shifting our understanding of human bodies, disease and experience (Biosocial Humans); and the opening up of new social, technological and emancipatory human possibilities (Human Futures).

Knowing humans

The COVID-19 crisis has thrown a spotlight onto the life-and-death stakes attached to how humans are known. This section of the book unpacks some of these issues, tracing how the choices made by policymakers, scientists and media are affecting how we are thinking about and experiencing the pandemic. *Ballo and Pearce* scrutinize how scientific knowledge has informed UK COVID-19 policy, and in particular the central role of modelling. They argue that models have illuminated tensions and shortcomings in the UK's relationship between science and politics. In response, Ballo and Pearce advocate for scientific models to become 'public objects', helping to reduce omissions and biases, as well as fostering public debate about the values underpinning pandemic policies. *Nerlich* analyses how metaphors have been central to the development of public knowledge about COVID-19 science and policy, and to making certain people seem more or less human. UK and US leaders are shown to favour metaphors of conflict and war, casting COVID-19 as an 'invisible enemy', while others prefer to use metaphors of collaboration and teamwork to emphasize the importance of collective action. However metaphors are used, Nerlich provides an urgent reminder of the need to remain alert to their meaning and consequences.

Moving from text to images, *Camilla Mørk Røstvik, Helen Kennedy, Giorgia Aiello and C.W. Anderson* analyse the role of 'everyday visuals' in knowing humans during the pandemic. They first identify and analyse three visuals that have become iconic representations of COVID-19: the virus, the 'flatten the curve' infographic and the face mask. The authors argue that while important, these iconic images do not tell the full story of COVID-19's visual language for two key reasons. First, there are many other generic visuals which stand in for the pandemic, such as empty public spaces and bar charts. Second, COVID-19 visuals can appear 'objective' yet omit tricky material such as death and inequality. *Vicari and Wang* use a Wuhan social worker's online diary to think through the role of place and platform in how societies have come to know COVID-19. Their comparison of Weibo and Twitter illuminates how these platforms have values embedded within them, informed by the political economies of China and the US respectively, which users work with, or sometimes against, in order to make their voices heard. Finally, *Cuevas-Garcia* also focuses on place, with a very personal account of how COVID-19 wrecked his wedding, planned for March 2020. The unravelling of plans exposes the tight coupling of apparently disparate actors, and how a single event can quickly erase old certainties. In an unfamiliar state of civic dislocation, Cuevas-Garcia's prospective guests were forced to confront not only shifting regulations regarding international travel, but also negotiate a new ethics of human mobility.

Marginalized humans

This section asks a key question related to difference, exclusion, marginality and disposability: how are different kinds of humans included or excluded by the contemporary moment of COVID-19? A range of policies, science advice and services during COVID-19 have had a major impact on the lives of marginalized people. Paying particular attention to disability, this section explores the processes of marginalization, how different forms of exclusion and inequality intersect, and how these processes are being resisted and transformed. *Goodley* sets the scene by exploring the extent to which researchers can and should engage with intersectional politics as a response to the new forms of exclusion shored up by COVID-19. Exploring how the routine devaluation and dehumanization of Black, poor, displaced and disabled lives has inevitably been exacerbated during COVID-19, *Goodley and Runswick-Cole* centre Rosi Braidotti's (2019) 'Missing People's Posthumanities' in order to make politicized calls for theory and practice that respond to the intersection of exclusionary processes of dehumanization. *Liddiard and the Co-Researcher Collective* demarcate the ways in which the global pandemic has heightened the worst effects of ableism and disablism in their lives. They reflect on the *new* vulnerabilities of COVID-19, the emerging labours around the management of risk and threat, and the affective realities in which disabled people are rendered disposable in times of crisis. *Pandey and Sharma* engage with the question of scientific advice and its publics, exploring the ways in which it is imagined and enacted, questioning the realities of who it aims to protect. Looking specifically at the multiple and intersecting impacts of lockdown, a severe migrant crisis and unending economic devastation in India, they offer a careful, critical analysis of the science and policy response to the COVID-19 crisis. *Titchkosky* focuses on political satire and what she labels 'the hierarchies of the human' during COVID-19. She considers how political pandemic satire both addresses and remakes divisions in society – along the same lines of precarity, disposability, marginality and intersectionality explored throughout this section. Importantly, Titchkosky's contribution examines how disability is relied upon to shame governments and the general public into improving the ways they are addressing the COVID-19 pandemic. However, this reveals the complex ways in which such satire buffers society against change by remaking the degradation of some people as a defining feature of humanity and its pleasures.

Biosocial humans

This section is concerned with how biological and biomedical knowledge shapes our understandings of what it means to be human. The contemporary era is marked by the production of many forms of new knowledge which both challenge and reinforce established notions of the human. *Martin* examines the impact of the rise of genomics in UK healthcare on how we understand the human. He analyses how the creation of massive gene-sequencing infrastructures linked to digital health records, and their application in testing and screening the population, is strengthening biological rather than social narratives about human health. In their chapter, *Goodwin, Shelton and Weiner* focus on the idea of human frailty, a concept that became increasingly important in the management of the care of elderly people during the pandemic. Their study examines how frailty has become conceptualized and operationalized in clinical decision making and policy, and the consequences of this for the treatment of elderly patients and our understanding of the ageing human. In contrast, *Williams'* chapter looks at a very different aspect of biomedicine. Her study of voluntary drives to collect biological samples that can be tissue-typed and placed on a stem cell donor registry illuminates the ways in which human communities are created, and how the pandemic has disrupted collective efforts to address racial inequalities. In a more personal reflection on the experience of blindness and ageing during the pandemic, *Michalko* gives testimony to the way in which COVID-19 has strengthened the power of sight as the 'master sense' and diminished the place of touch in an increasingly disconnected online world. He also questions ideas of care and the role of science in the control of human lives.

Human futures

Early on in the pandemic, two things happened which had been until then unthinkable: a global shutdown of economic activity and a global outpouring of 'we feeling', manifesting not just in the claims that 'we are all in this together', but in the sudden rise (and very fast artistic uptake) of new platforms for engagement, of solidarity with and interest in the progression of the virus in distant countries. This simultaneously involved the shrinking of the world and an expanding cornucopia of new possibilities. As the pandemic has progressed, things hardly look that hopeful anymore. Different countries now cycle through various stages of lockdown, vaccine dissemination, and a determined resurgence of old ways of doing things, whether realistic or even desirable. The four chapters in this section are less about what the virus has been, and more about what it teaches us about being human in the face of an increasingly uncertain future. This is a future in which different kinds of humans might flourish together, but it may also be one of social and ecological breakdown. Will we genetically edit the next generation to be COVID-19 resistant? Can we really rethink how the economy works or how we educate our children? In this section the authors consider not so much *what* will be, but how we imagine what *could* be, examining the future as something co-created between science, technology, politics and society in the here and now. It begins with a collective response by *Arthur, Amanatidou, de Saille, Birabi and Pandey* which considers how the notion of 'responsible stagnation', as a way of coping with long-term economic slowdown, could help shift our focus from an economy based on outcomes and GDP-measured productivity to an economy of processes focused on wellbeing. *Morrison* then reflects on what COVID-19 tells us about the perils of heritable genome editing. *Pilson* explores how we might use the lessons taught to us by the crisis to reimagine inclusive education for post-COVID-19 times. Finally, *Raven* examines how the arts, humanities and social sciences might use COVID-19 to reassert their claim on futurity, to rehabilitate utopian praxis, and to counter the solution-focused paradigms of technoscience and the associated neoliberal catechism that 'there is no alternative' to the way we have lived until now.

In the concluding section, the editors draw together the many strands touched on by the 17 chapters, and weave together a summary that reflects on what it means to be human during and, possibly, after COVID-19.

About iHuman: disruptive research into what it means to be human

The Institute for the Study of the Human (iHuman) was established in 2018 in the Faculty of Social Sciences, University of Sheffield, to develop innovative multidisciplinary research on what it means to be human in the 21st century. Our approach combines work at the interface of critical disability studies (CDS) and science and technology studies (STS), together with a wide range of other disciplines in the social sciences, arts and humanities, natural sciences, engineering and medicine. We aim to create a safe space for genuine cross-disciplinary conversations and promote experimental methods, theories and new ways of working together. In addition, we are deeply committed to building a local, national and international community of scholars interested in the human in all its forms as a subject of analysis. At the heart of our work is a commitment to promoting equality and diversity, developing the next generation of scholars through training and supporting career development, and achieving engagement with a wide range of communities to achieve real world change. For more information go to: https://www. sheffield.ac.uk/ihuman.

PART I

Knowing Humans

The COVID-19 crisis has thrown a spotlight onto the life-and-death stakes attached to how we humans know about ourselves, and what we know other humans to be. One might argue that the pandemic has recreated epistemological anxieties, ontological uncertainties, and methodological divisions. This section of the book will unpack some of these issues, tracing how the images, metaphors and models used by scientists and media, as well as the choices made by policymakers, are affecting how we are thinking about and experiencing the pandemic.

ONE

Making Models into Public Objects

Rokia Ballo and Warren Pearce

When the UK government switched its COVID-19 strategy from mitigation to suppression on 17 March, it seemed a case of 'new data, new policy'. News reports cited how modelling from Imperial College's Centre for Global Infectious Disease projected between 410,000–500,000 deaths, prompting the government to implement stringent lockdown policies, and upending the UK's political economy. Unsurprisingly, given the stakes involved, public debates about government policies and their effects have been intense; for example, the time taken for the UK to introduce a lockdown policy and the unequal effects on health and wellbeing. However, two broader questions regarding the relationship between science, politics and publics now demand reflection: how is science advice being formulated, and how is that advice being used to justify policy decisions in the public interest? We attempt to answer these questions by focusing on whether the models that lie at the heart of COVID-19 science can become 'public objects', helping to foster both public debate about policy and trust in the institutions that we rely on to make life-and-death decisions.

What's missing from evidence-based policy?

Despite the UK government's recent rocky relationship with experts, scientific advisers were placed at the forefront of the public response to COVID-19, both rhetorically, through phrases such as 'following the science', and visually, with the Prime Minister being flanked by scientific advisers at daily press conferences. The government and its experts appeared to be working harmoniously on 'evidence-based policy', the rational ideal of creating knowledge for decision making. However, relying on science for the public validation of policy conceals an array of judgements about what that science includes and excludes, and what this in turn says about how, and for whom, the state cares.

In the early days of the pandemic, epidemiological modelling dominated UK science advice. On the surface, the government's response to the Imperial College report seemed like the ideal of how evidence-based policy should work. Yet, if this modelling was decisive in prompting policy change, then it should also be acknowledged that models imagine human experience as a rather homogeneous affair. Keeping a model simple can be useful in demonstrating a particular dynamic that is relevant to a policy question; for example, Imperial College modelling provided charts and curves which predicted the future, based upon a number of largely fixed assumptions. The problem comes when the omissions from model simplifications become naturalized as part of political decision making, fostering an ignorance of less quantifiable issues which thus risk being excluded from consideration.

Science and politics in pandemic modelling

The tensions between science and politics, that have received such sustained attention during the pandemic, have long been studied by scholars of science and technology studies (STS). For example, Roger Pielke Jr's influential idea of the 'honest broker' (2007) posits the science advisor as providing a broad range of options for policymakers to choose from, an approach recognizable within the UK system of Chief Scientific Advisers. However, as Sheila Jasanoff (2008) suggests, society may not always be best served by advisors providing a broad range of policy options, and that in some, often value-laden, situations, scientists have an important role in helping to build consensus around policy. Imperial College's modelling had an important role in this regard, inviting more urgent and stringent policy interventions than had previously been contemplated. Yet this central role for experts also presents a democratic challenge. As Jasanoff (2008) recognizes, experts' values are always embedded within processes of data collection and processing such as modelling, so must be accountable not only to the scientific community but also to wider society. This has proved controversial within COVID-19, as a government struggling to keep pace with the rapidly evolving crisis has foregone parliamentary processes, raising concerns about increasingly unchecked state power. Whether scientific advisory bodies such as the Scientific Advisory Group for Emergencies (SAGE) act as honest brokers or more forceful policy advocates, the legitimacy of experts to convey what would best serve public needs remains questionable when publics are given scant opportunity to represent their own interests (Boschele, 2020).

Science is a deeply human activity, embedded with the same assumptions and biases that help maintain social inequalities. Models are undoubtedly useful aids to policymakers, but are in no way immune from perpetuating existing patterns of bias and exclusion. For example, the minutes of SAGE meeting 14 show how early discussions of lockdown policies were shaped by assumptions of public non-compliance, which in turn influenced how policy options were presented, with one key adviser later stating that it was hard to imagine how compliant people would be with a stringent lockdown. This scientific assumption of the typical British citizen as prioritizing personal freedom over collective benefit, echoing the Prime Minister's description of the British people as 'freedom-loving', helped to delay lockdown, a decision later admitted by some advisers to have increased the UK's death toll. More broadly, while lockdown's epidemiological benefits were eventually modelled and accepted by the government, some wider implications of the policy were not prominent in expert advice. For example, lockdown had grave implications for those for whom home is not a safe or permanent place, with the Office for National Statistics reporting a rise in demand for domestic abuse victim services in 2020.

In pointing out the questionable assumptions that underpinned an apparently evidence-based approach, we do not advocate for evidence from models to be rejected wholesale. Rather, we caution against overreliance on the pictures of society presented by models, particularly as the democratic power of publics to reject policies that do not serve their interests has been eroded during the pandemic. Instead, we wish to consider how making models more public objects, through greater public involvement in their development and use, might alert scientists and ministers to their deficiencies and support a more equitable prioritization of the variables included in models.

Inequalities, biases and pandemic modelling

Although the pandemic has been felt by all, its effects have not been felt equally across society. The inequalities fuelling differential experiences of the pandemic are complex, far-reaching and long-standing. Understanding COVID-19 's social impact is now recognized as important for the UK's recovery, but this was not always the case. When well-known links between inequalities and poor health re-emerged in hospital data, existing biases went unchallenged as the government failed to act.

In April, the government advisory group, COVID-19 Clinical Information Network (CO-CIN), began investigating the disparities in COVID-19 outcomes across ethnic groups. Their final report confirmed the disproportionate impact of COVID-19 among Black and minority ethnic (BAME) patients, suggesting that a range of factors, including comorbidities and social deprivation, which are consistently higher in BAME populations, could be the cause. Yet publicly, the government chose to continue pushing the narrative of COVID-19 as an indiscriminate equalizer during daily briefings: particularly following the Prime Minister's own hospitalization that month. Minutes from SAGE 29 further reveal the government's unwillingness to acknowledge inequalities such as deprivation as important, focusing instead on comorbidities, namely diabetes, as the sole explanation for unequal outcomes. Research has shown how health inequalities are often blamed on patients' lifestyle choices, while also confirming that comorbidities, like Type 2 Diabetes Mellitus, are strong indicators for broader inequalities sufferers may be living with – suggesting both systemic and individual causes. The government's decision to use comorbidities to explain the disproportionate impact of COVID-19 on marginalized groups reflects these preexisting biases against health inequalities, and a tendency to blame individuals rather than address the links between issues like deprivation and poor health outcomes, even in the face of scientific data. Intended or not, this approach legitimized the presence of unequal COVID-19 outcomes rather than attempting to address them. To date no direct political action to address the exacerbation of inequalities has emerged, even following Public Health England's disparities report published in June 2020, as well as a distinct absence of inequalities as a variable for models to account for.

At this stage, we cannot be certain why these decisions were made. However, with the information available we make two observations, regarding government ministers' apparent unwillingness to explore the relevance of inequality, and the impact a more public science may have had in such circumstances. First, if COVID-19 impacts on marginalized groups being more than comorbidity-driven were taken more seriously, would a policy that targeted marginalized groups, particularly on ethnic grounds, have been accepted as within the bounds of political possibility? We suggest not, which – whether or not it was the 'right' policy – may have dissuaded scientific advisers working 'with the grain' of politicians from pursuing it (Palmer et al, 2019). Second, any potential remedy to this does not lie in an attempt to artificially separate science and politics in the science advice process. Biases are inevitable, particularly in narrowly-constructed groups of science advisors where representation of marginalized groups remains woeful. The challenge is to mitigate them through transparency and public discussion that allows a wider range of expertise to contribute to science advice, including the valuable knowledge a diverse range of publics could bring. This kind of openness could help prevent inequalities from being an afterthought, while also subjecting scientific evidence for policy to greater democratic accountability (Pearce, 2020).

Making models into public objects

COVID-19 has shown models to be valuable tools for producing scientific knowledge claims for the development of evidence-based policy. The pandemic has also shown that the intertwining of science and politics is only set to increase, moving forward. We suggest that as models become an ever more important part of UK policymaking, we must openly discuss their limitations as well as their potential. Our critique is not a repeat of current, high-profile discussions about how to increase public trust in science. We suggest that such questions are missing the point, falling into old patterns of positioning publics as secondary to science, rather than a more cosmopolitan approach to knowledge focused on how we make science *worth* trusting to begin with, and who gets to define what constitutes the public good (Raman and Pearce, 2020). If models are only as good as the questions asked of them, then we need to consider which problems models will or will not be asked to solve, who decides this, and how accountability can be strengthened through public inclusion in the modelling process.

One solution is for models to be treated as public objects. This means a deeper understanding of 'public' than just increasing transparency, important as that is. Rather, it means thinking of models as a means to create much-needed space to discuss public values, concerns and needs. For those concerned about the status of science in democracies, we suggest that this is a more fruitful way forward than a narrow focus on 'trust in science'. When the assumptions and deficiencies of scientific modelling are a matter of life-and-death, appeals to scientific authority can go awry. Maintaining science's role in democracy means being comfortable, both with the likely ambiguity of scientific knowledge, and how science and human values must be knitted together when debating and defining what constitutes the public good.

References

Boschele, M. (2020) 'COVID-19 science policy, experts, and publics: why epistemic democracy matters in ecological crises', *OMICS: A Journal of Integrative Biology*, 24(8): 479–82.

Jasanoff, S. (2008) 'Speaking honestly to power', *American Scientist,* 96(3): 240–3.

Palmer, J., Owens, S. and Doubleday, R. (2019) 'Perfecting the "elevator pitch"? expert advice as locally-situated boundary work', *Science and Public Policy* 46(2): 244–53.

Pearce, W. (2020) 'Trouble in the trough: how uncertainties were downplayed in the UK's science advice on COVID-19', *Humanities and Social Sciences Communications* 7(1): 1–6.

Pielke, Jr., R.A. (2007) *The Honest Broker: Making Sense of Science in Policy and Politics*, Cambridge: Cambridge University Press.

Raman, S. and Pearce, W. (2020) 'Learning the lessons of climategate: a cosmopolitan moment in the public life of climate science', *WIREs Climate Change* 11(6): e672.

TWO

Pandemics, Metaphors and What It Means to Be Human

Brigitte Nerlich

Knowledge and human knowledge

Knowledge and metaphor go hand in hand. We create and expand knowledge through metaphors and we need knowledge to understand metaphors. Metaphors are crucial to the production of knowledge, as they allow humans to make connections between what is already known and thinkable and what is not yet known and neither thinkable nor sayable. By mapping meaning from one domain of knowledge or perception onto another, metaphors create new meaning. This includes knowledge of what it means to be human.

The ability of metaphor to make us see things in certain ways, and from there to understand and explain them, is fundamental to seeing, understanding and dealing with unknown and invisible phenomena. At the end of 2019 a new virus, severe acute respiratory syndrome coronavirus 2 (SARS-CoV-2) emerged in China, with the World Health Organization declaring the associated disease, COVID-19, a pandemic on 11 March 2020. As early as 21 January, the US Centers for Disease Control asked a science illustrator to provide an image of the virus. This image of a spikey ball, and many others like it, spread around the world and became one of the first ways in which people came to know the virus and the pandemic.

The spikey ball inspired numerous subsequent images, as well as many polemical takes on how politicians had failed to deal with the virus. This chapter outlines how two heavily criticized leaders, former US President Donald Trump and UK Prime Minister Boris Johnson, used metaphors to shape a verbal image of the virus that served their political purposes.

All knowledge is political but some is more political than the rest. Metaphors play an important role in bringing knowledge and politics together, but can enhance ignorance as well as understanding.

Metaphors that make us less human

In April 2020, former President Trump started to repeatedly use a familiar and worrying metaphor, describing the virus as the 'invisible enemy' (Chan, 2020): 'In light of the attack from the Invisible Enemy, as well as the need to protect the jobs of our Great American Citizens, I will be signing an Executive Order to temporarily suspend immigration into the United States'.

Since the emergence of the germ theory of disease, bacteria and viruses have been framed as invisible enemies to fight, battle and wage war against. This ubiquitous metaphor is closely linked to what some have called 'biomilitarism' (Montgomery, 1996): a framing where we 'just know' that germs have to be 'fought', despite not all germs being bad for us. Out of the mouth of Trump this metaphor of invisible enemies took on a more sinister meaning, repeatedly framing immigrants as infestations and disease, for example in a tweet from 18 June 2018, where he wrote that immigrants 'pour into and infest our country'.

Such metaphors have a long and sinister history, and shape who we think we are and who or what others are. The use of 'invisible enemy' for the virus, and the fact that Trump linked it specifically to immigration, continues this infamous framing, which turns immigrants into vermin or virus and virus into immigrants. Jack Shafer wrote an illuminating analysis of Trump's use of the 'invisible enemy' metaphor for *Politico*. He pointed out that: 'Trump's determination to label the virus an invisible enemy bears all the hallmarks of a branding campaign, one fashioned to shape our attitudes toward the microbe to his liking. By calling the virus 'invisible', Trump implies that he can't be responsible for its wreckage because who can be expected to see an invisible thing coming? And once the unseeable thing has arrived, there are limits to what one can be expected to do about it!' (Shafer, 2020). This means that once we know the virus as an invisible enemy, we also know that we are not responsible for its actions. Trump also sought to distance himself from the virus by calling it the 'Chinese virus' (in March) and even the 'China plague' (in June during the Black Lives Matter protests), again othering people who are not like him and his followers. The impacts of this implicit racism quickly became apparent, as racist acts and harassment against Asian people increased in the US and elsewhere.

This use of metaphors tells us not only something about how a certain politician sees and conceptualizes the virus, but also about how he sees and conceptualizes humans. To quote Shafer again: 'Trump's determination to brand the virus as an enemy, rather than a pathogen, pays political benefits. Calling his crusade "our big war" and directly enlisting the military in the fight allows Trump to frame a public health crisis as a military operation: he is the commander in chief, we are his foot soldiers, our patriotic duty is to obey him, and the entire planet is his battleground' (Shafer, 2020). As would become clear during the June 2020 Black Lives Matter rebellion, what mattered to Trump was 'domination', not only of viruses, but of humans, especially humans that are other than him, such as women, immigrants, people of colour, and people with health conditions.

What about Trump's counterpart in the UK? During his first speech since recovering from COVID-19, on 27 April, Prime

Minister Boris Johnson similarly emphasized the invisibility of the virus, describing it as an 'unexpected and invisible mugger', providing a twist on Trump's martial metaphor (https://www.gov.uk/government/speeches/pm-statement-in-downing-street-27-april-2020). The mugger metaphor framed the virus as providing an unexpected and unseen attack, attempting to counter criticisms that the UK government was, in fact, underprepared for an eventuality that could and should have been predicted. As David Shariatmadari pointed out: 'the prime minister's idea that we "wrestle [coronavirus] to the floor" would seem at odds with the patient, precise work that will have to be done, over many months, to keep it at bay' (Shariatmadari, 2020).

Both Trump and Johnson fell ill with COVID-19. Both were praised for their 'fighting spirit'. As Ed Yong, one of the most competent, indeed Pulitzer Prize winning, pandemic communicators has pointed out in one of his many articles for *The Atlantic* (2020): 'Equating disease with warfare, and recovery with strength, means that death and disability are linked to failure and weakness'; a nefarious framing reinforced in July 2021 by Sajid Javid, the UK Health Secretary, who suggested people should not 'cower' from the virus, after he himself had recovered from it.

War metaphors can be used to rally people, in a sense to 'mobilize' them; they can even create solidarity within some groups of people. However, behind their bullishness they can hide the fact that patient, precise and well-planned work is not being carried out by governments. They also divert attention away from government responsibilities and focus instead on individual responsibilities, individuals who are asked to hand-wash and social distance, to not 'flood' parks and beaches or crowd together in protest marches. It diverts attention from the impacts of austerity, from the critically underfunded healthcare system, and from the hostile environment for migrants and people of colour. Most importantly it diverts attention away from our common humanity.

Metaphors that make us more human

War metaphors sometimes appear so ubiquitous, it seems impossible to imagine alternatives to battles, fights and victories. Yet the pandemic has demonstrated how other ways of seeing and understanding COVID-19 are possible. For example, in Germany and New Zealand, (then) leaders Angela Merkel and Jacinda Ardern have refrained from war metaphors, focusing instead on supportive and compassionate policies harnessing community, collaboration and teamwork. Ardern has regularly referred to New Zealanders as a 'team of five million' in an effort to unite people and encourage them to follow her government's advice to curb the virus' spread (Flanagan, 2020). The same 'unite' message has also been used during New Zealand's recent vaccination campaign.

Merkel, in turn, avoided metaphors almost completely, and stuck to explaining the science behind the pandemic and behind the policies to deal with it. 'Her first major public interaction during the crisis was a televised address on 18 March. Merkel's words to describe the crisis were simple and straightforward. She spoke of "this situation", "a historical task", and a "great challenge" ahead' (Paulus, 2020). Merkel, Ardern and other female leaders around the world seem to have avoided authoritarian and denigrating metaphors. This might just be a coincidence and might be related to many other factors apart from gender, but recent research has highlighted the 'centrality of misogyny in legitimating the political goals and regimes of a set of leaders in contemporary democracies' (Kaul, 2021) – leaders that are all male.

However, it is important to stress that metaphors used by those communicating the pandemic on the ground contrast starkly with divisive war metaphors used by some politicians. They are more tied in with science and health communication and are both explanatory and support communal action. Examples are the train and football metaphors used by the UK's then Deputy Chief Medical Officer Professor Jonathan

Van-Tam to explain the spread of the virus and the policies to deal with it, and wall metaphors used by the Chief Medical Officer Professor Chris Whitty to stress how important it is to get vaccinated in order to build a community barrier against the virus. These are good ways to communicate emerging science and emerging policy interventions. The metaphors also created new common knowledge of the virus and the measures to deal with it, namely vaccines. I absolutely agree with Van-Tam who told the BBC: "I love metaphors. I think they bring complex stories to life for people" (Morton, 2020). I think Van-Tam speaks for many science communicators here.

Many verbal and visual examples of non-war metaphors can be found in a collection crowdsourced during the pandemic called #ReframeCOVID-19, a collective initiative launched to promote non-war-related language on COVID-19. An ecological alternative to war metaphors can be found in an article by Michael Hanne (Hanne, forthcoming).

Conclusion

Metaphors are indispensable for creating and expanding knowledge, but they can also twist and distort human understanding and the understanding of what it means to be human. Let's celebrate metaphors that bring knowledge and people together, and shine a critical light on those metaphors that drive people apart and destroy common and communal knowledge and understanding. Among all the anxiety and destruction there is also hope. As the Public Interest Research Centre has pointed out, after dissecting and rejecting some of the war metaphors used in the crisis: 'The shared experience and interconnectedness of being human has been brought sharply into focus. Our health is inextricably linked to the health of our neighbours. Our resilience is community resilience. In the face of this crisis to cooperate and collaborate is not a choice, it is the only way to respond.' (Sanderson and Meade, 2020). Let us not be distracted from this communal action by politicians wielding divisive and derogatory metaphors, and let's celebrate those that offer metaphors that enable people to understand the pandemic and act in ways that mitigate its impacts.

References

Chan, A. (2020) 'Trump to temporarily suspend immigration into US amid pandemic, cites "invisible enemy"'. *International Business Times*, 20 April, https://www.ibtimes.com/trump-temporarily-suspend-immigration-us-amid-pandemic-cites-invisible-enemy-2962091

Flanagan, M. (2020) 'Lead with compassion and ditch the wartime metaphors', The Article, 7 May, https://www.thearticle.com/lead-with-compassion-and-ditch-the-wartime-metaphors

Hanne, M. (forthcoming) 'How we escape capture by the "war" metaphor for COVID-19', *Metaphor and Symbol*.

Kaul, N. (2021) 'The misogyny of authoritarians in contemporary democracies', *International Studies Review*, 23(4): 1619–45, https://doi.org/10.1093/isr/viab028

Montgomery, S.C. (1996) *The Scientific Voice*, New York: Guilford Press.

Morton, B. (2020) 'Jonathan Van-Tam's best analogies: penalties, equalisers and yoghurts', *BBC News*, 3 December, https://www.bbc.co.uk/news/uk-55169801

Paulus, D. (2020) 'How politicians talk about coronavirus in Germany, where war metaphors are avoided', *The Conversation*, 22 May, https://theconversation.com/how-politicians-talk-about-coronavirus-in-germany-where-war-metaphors-are-avoided-137427

Sanderson, B. and Meade, D. (2020) 'Unimaginable times', *Public Interest Research Centre*, 20 March, https://publicinterest.org.uk/part-1-unimaginable-times/#more-11028

Shafer, J. (2020) 'Behind Trump's strange "invisible enemy" rhetoric', Politico, 4 September, https://www.politico.com/news/magazine/2020/04/09/trump-coronavirus-invisible-enemy-177894

Shariatmadari, D. (2020) 'Muggers and invisible enemies: how Boris Johnson's language hints at his thinking', The Guardian, 27 April, https://www.theguardian.com/politics/2020/apr/27/muggers-and-invisible-enemies-how-boris-johnsons-metaphors-reveals-his-thinking

Yong, E. (2020) 'What strength really means when you're sick', The Atlantic, 9 October, https://www.theatlantic.com/health/archive/2020/10/trump-strength-coronavirus/616682/

THREE

The Role of Everyday Visuals in 'Knowing Humans' During COVID-19

Camilla Mørk Røstvik, Helen Kennedy, Giorgia Aiello and C.W. Anderson

The human condition during COVID-19 has been communicated through a barrage of news stories about the pandemic, and much of that news has been visual. Line charts, bar charts and data visualizations have become key to public communication about the pandemic. Likewise, photographs are also used to represent the virus and illustrate the pandemic's impact on aspects of everyday life, like working, going to school or socializing at the pub. Writing in 2020, Julia Sonnevend argued that visual representations of COVID-19 'are entry points for public discussion and social contestation' (2020: 452) and, as such, would play an important role in persuading people to act responsibly. 'Without the ability to see and relate to this crisis', she wrote, 'people will be unlikely to follow strict guidelines that interfere with their usual daily lives' (Sonnevend, 2020: 452). In short, visuals have been ubiquitous and pivotal in COVID-19 communication. In this chapter, we discuss three visuals which have become iconic of the pandemic: images of the virus itself, the flatten-the-curve-line graph, and the face mask. Then, unlike Sonnevend and other writers who prioritize iconic images like these, we propose that it is also important to attend to the more generic everyday visuals of COVID-19 that circulate.

Among those visuals that have become representative of the pandemic is *the ubiquitous, computer-generated image of the virus itself* (Figure 3.1). In January 2020, journalists, editors, scientists and politicians needed an image that would convey COVID-19's seriousness. When the US Centers for Disease Control and Prevention (CDC) realized how serious COVID-19 would become, one of their first public relations moves was to ask medical illustrator Dan Higgins to give the disease a 'visual identity'. The resulting 3D image was based on scientific data, but artistic licence was also involved: the colours red, orange and yellow were chosen to alarm the public and shadows were added to make it seem more real (Fairs, 2020). Released on 31 January 2020, the image circulated around the world. It was and continues to be used in news and on social media, sometimes

Figure 3.1: Coronavirus illustration

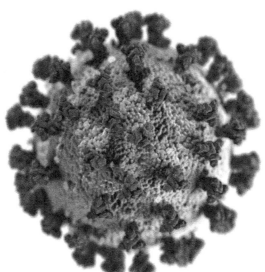

Source: https://www.cdc.gov/media/subtopic/images.htm

with tweaks to avoid copyright issues, although there were in fact none, as the CDC wanted the image to spread. By the end of the year, the image had come to visually signify the COVID-19 crisis. Sonnevend (2020) describes this ubiquitous image as an aesthetically pleasing 'close-up' that synthesises key information about the virus while also being versatile, thanks to its economy of detail and three-dimensional form. Indeed, it has been widely adapted, in abstract, Art Deco, cartoon, and other forms, in montages and stock photography, and as symbols in data visualizations. It represents the disease and as such has become iconic.

The *'flatten-the-curve'* infographic is another ubiquitous COVID-19 image (Figure 3.2). It is a line chart with two lines: the first, sharp curve represents the pandemic outbreak without intervention, and the second is a curve flattened by intervention measures such as social distancing and lockdowns. It is based on a model from earlier epidemics and pandemics like the 1918 Influenza and SARS in 2002, which is widely

Figure 3.2: Flatten the curve infographic

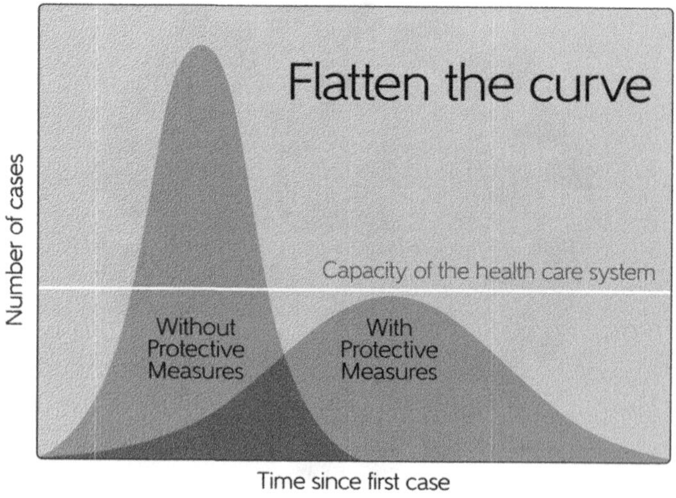

Source: Shutterstock

used in epidemiological circles (Cotgreave, 2020). A digital version had been used by scientists in 2007, for example, to demonstrate various scenarios in relation to pandemic strategic planning for the future. Data visualiser Rosamund Pearce based her original COVID-19 flatten-the-curve visual for an article in *The Economist* on this existing model (The Economist, 2020). Like the original, Pearce's model was not based on actual data. Rather, it aimed to illustrate the overall problem and solution. Population Health Professor Drew Harris was later inspired to alter the visual by adding a horizontal line showing 'healthcare capacity'. This was a 'fully theoretical invention' which intended to communicate the message that interventions were needed to avoid healthcare systems breaking down (Wilson, 2020). The image went viral and variations abounded, so much so that data visualiser Andy Kirk joked on Twitter that we need to 'flatten the curve of new versions of the flatten the curve chart' (https://twitter.com/visualisingd ata/status/1242420727401394176?s=20).

The face mask is a third image that has come to visually represent the pandemic. Images of people wearing *face masks* have been ubiquitous during COVID-19, their role shifting over time, from a garment associated with medical professionals, to a protective measure in East Asia, to their everyday usage around the world. Images of mask-wearers underpinned mainstream political advice in most parts of the world to wear masks for protection. In contrast, images of non-mask-wearers showed rule-breakers, like those who broke into the US Senate in January 2021 or anti-lockdown protesters around the world. Some photos of mask-wearers came from stock photography suppliers like Getty Images, whereas some accompanied specific, human interest stories, such as 90-year-old Freda France from Sheffield, depicted on a BBC Yorkshire webpage wearing a tiara and encouraging people to get vaccinated (Figure 3.3). In addition, the mask has frequently been used as an illustration of measures that the public was required to adopt to reduce the spread of the virus,

Figure 3.3: Freda France wearing a face mask

Source: BBC Yorkshire, https://www.bbc.co.uk/news/
uk-england-south-yorkshire-55321177

for example in the UK government's 'hands – face – space' campaign in the middle of the pandemic. Thus the meaning of the mask image changed frequently, yet it continued to look the same, whether in a photograph or as an illustration.

These visual representations of the pandemic – the virus molecule, the flatten-the-curve graphic, and images of face masks – have become iconic. The face mask image, for example, became an icon of how people's appearance and everyday lives were modified during the pandemic.

Yet pandemic visuals extend far beyond these iconic images. What we describe as 'generic visuals' – standardized in appearance, performing particular design functions, and widely circulated in the news media – also circulate widely. Generic COVID-19 visuals include the many photos of nameless mask-wearers and of public spaces left empty by the pandemic, myriad daily bar charts, line charts and coloured maps of infection rates, vaccination rates, and changing disease control measures, and other image types, some of which combine visual elements from both photos and graphs. Generic visuals increasingly populate journalism and other information sources, online and off, on mobile apps and other digital platforms. They are what Frosh (2003) calls the visual 'wallpaper' of our lives, suggesting we may have come to consider them as meaningless decoration. They are also the visual wallpaper of the COVID-19 crisis. Generic visuals are much more ubiquitous than iconic images yet, surprisingly, research on news visuals – which is not commonplace – tends to examine iconic images, such as arresting photographs or award-winning data visualizations. Despite our increasing exposure to generic visuals, we know very little about how they contribute to framing issues and events in and through the news. We are filling this gap with our research project: Generic Visuals in the News (https://genericvisuals.leeds.ac.uk/).

Researching the myriad of everyday, generic images of COVID-19 makes it possible to reflect on what is missing from news visuals of the pandemic. Visualizing COVID-19 offers a way of grasping a fast-evolving event and making it knowable. And yet, centred as they are on visual conventions that stylize, abstract or decontextualize, generic visuals cloud important aspects of the pandemic, such as its bodily and material impacts, or, put more brutally, illness and death. In a New York Times opinion piece, Sarah Lewis (2020) asks: where are the photos of the people dying of COVID-19? She argues that 'we're not seeing this crisis with our own eyes' and that this has consequences in terms of *not* mobilizing people to act. She goes on to say that 'statistics alone, however clear, are not historically how we have communicated calamity on this scale'. In her view, a particular type of image, photographs, are needed to move people to act.

Data visualizer Mona Chalabi has a different view on the problematic absences in visual representations of the pandemic. Early in the crisis, she highlighted the absence of visual representations of the inequalities that COVID-19 exacerbates (Chalabi, 2020). In response, she attempted to address this gap through her own work, visualizing, among other things, who has the privilege of being able to work from home, and the ways in which the virus disproportionately affects Black Americans. Chalabi's work points to the fact that choices are made about what to represent visually, what not to visualize, and how to visualize, all of which tell us something about the human condition during the pandemic, and the politics thereof.

This is another reason why it is important to attend to the role of everyday visuals in 'knowing humans' during the pandemic. The forms and affordances of photos, data visualizations, and other image types combine to communicate a sense of facticity or objectivity, and yet tricky material is missing, like death and inequality. Everyday visuals of COVID-19 do different things. As we argue elsewhere (Aiello et al, Forthcoming), they convey a sense of nationalism, but also of localism and cosmopolitanism. The ubiquity of steep curves in line graphs and of scary molecules means these images have both iconic and generic characteristics – they are arresting and memorable, yet they are standardised and circulate with frequency. Photos of vaccinations and charts of vaccination rates may reassure, and all of these image types may be seen as meaningless visual wallpaper, or they may assemble publics and mobilize them to act (Aiello et al, Forthcoming). These possibilities, and the slippage between the iconic and the generic, are further reasons to focus on the role that generic visuals play in communicating COVID-19. In other words, and to conclude, we might make a better, more human future by understanding the role that generic visuals play in knowing humans, in the context of COVID-19 and more broadly.

References

Aiello, G., Kennedy, H., Anderson, C.W. and Mørk Røstvik, C. (Forthcoming) ' "Generic visuals" of COVID-19 in the news: invoking banal belonging through symbolic reiteration', *International Journal of Cultural Studies*.

Chalabi, M. (2020) 'To understand inequality in New York', *Twitter*, 11 June, https://twitter.com/MonaChalabi/status/1271100297037918209/photo/1

Cotgreave, A. (2020) 'Visualizing COVID-19: a discussion on the "flatten the curve" visualization and responsible data use', *Tableau*, 27 March, https://www.tableau.com/about/blog/2020/3/covid-19-discussion-history-flatten-curve-visualization

Fairs, M. (2020) 'Iconic COVID-19 images designed to create 'a feeling of alarm' says CDC medical illustrator Dan Higgins', *Dezeen*, 14 May, https://www.dezeen.com/2020/05/14/covid-19-images-coronavirus-cdc-medical-illustrator-dan-higgins/

Frosh, P. (2003) *The Image Factory: Consumer Culture, Photography and the Visual Content Industry*, Oxford: Berg.

Lewis, S. (2020) 'Where are the photos of people dying of COVID-19?' *New York Times*, 1 May, https://www.nytimes.com/2020/05/01/opinion/coronavirus-photography.html

Sonnevend J. (2020) 'A virus as an icon: the 2020 pandemic in images', *American Journal of Cultural Sociology*, 8(3): 451–61.

The Economist (2020) 'COVID-19 is now in 50 countries, and things will get worse', *The Economist*, 7 February, https://www.economist.com/briefing/2020/02/29/covid-19-is-now-in-50-countries-and-things-will-get-worse

Wilson, M. (2020) 'The story behind "flatten the curve", the defining chart of the coronavirus', *Fast Company*, 13 March, https://www.fastcompany.com/90476143/the-story-behind-flatten-the-curve-the-defining-chart-of-the-coronavirus

FOUR

Humans, COVID-19 and Platform Societies

Stefania Vicari and Zheng Yang

In January 2020, Guo Jing's diary from quarantined Wuhan made it to BBC News' Twitter feeds (Figure 4.1). A social worker newly settled in the city; Jing narrated the first COVID-19 lockdown by sharing fragments of her lonely life. She wrote about her fears of life and death, often hinting at dysfunctions in the Chinese authorities' handling of the pandemic outbreak (BBC News, 2020a).

The first journal entries were published on a WeChat public account but Jing herself described the challenges she had to face when trying to live report the pandemic on Chinese social media:

> The first day when I tried to post my diary on Weibo, the photos didn't go through. The text of my diary didn't go through either. I had to convert the text into an image file to post it. Yesterday, even if I converted texts into an image file, I still failed to post it in my WeChat moments. After I posted it on Weibo, access to it was evidently limited. (Jing quoted in Yang, 2020)

The diary made it to BBC News, and to Twitter, because of the news value of its human – and humane – essence, an essence that, back then, sounded extremely 'exotic' to Western audiences. This 'exoticness', we now know, would not last very long.

Figure 4.1: BBC News tweet about Guo Jing's diary

Source: https://www.bbc.co.uk/news/world-asia-china-51276656

We take Guo Jing's story as a starting point to reflect on COVID-19, humans, and digital platforms, because it sheds light on the way these three entities differently intertwined in China and in the West at the start of the pandemic. We reflect on how platforms, in their context of use and with their specific affordances, shaped the way COVID-19 came to be known, understood and, ultimately, lived in China and in the West.

When we think of the functions, uses and values of social media platforms, we are often tempted to think of a universal 'digital society', primarily drawing on Western or Global North contexts and models (Chan, 2013). And yet, by picturing features of social media as somehow intrinsic in technology, we neglect to account for the way in which broader environments and contexts shape platform uses in everyday life (Willems, 2020). For instance, a digitally universalistic take would prevent us from understanding the unusual journey of Guo Jing's diary, from popular but also ostracized digital trace in the ecosystem of mainstream Chinese social media platforms, to key clickable source for legacy media accounts on Western social media.

To understand how COVID-19, humans and digital platforms differently intersected at the outbreak of the pandemic, we spent some time on the two leading mainstream microblogging platforms in China and in the West – Weibo and Twitter. We observed how emotions, themes and information sources differently emerged in COVID-19 content. Our aim was that of reflecting on the similarities and differences across the Western and the Chinese model of 'platform society' when it comes to host participatory dynamics. It would, of course, be reductionist to assume the existence of a monolithic 'Western model' that fully encompasses the different ways in which citizens from across Western contexts engage with public debates on social media. What we are presenting here is more an exploration of how COVID-19 and humans differently intersected in platform ecosystems predominantly shaped by state (China) or free market (West) capitalist values.

Microblogging the pandemic in different platform ecosystems

As limited as they may be in capturing nuanced dynamics in social media conversations, automated text-mining techniques can enable, among other things, the identification of themes and patterns within a large number of social media posts; for example, the use of polarizing language. We used sentiment analysis, semantic text mining and network mapping to explore COVID-19-focused content on Weibo and Twitter over a 48-hour period in the early phases of the pandemic in China (February 2021) and in the West (March 2021). We hoped this would give us an initial sense of the similarities and differences in the way social media users talked and learned about the events on the two platforms.

As a matter of fact, in early February, Weibo posts showed more polarised sentiment than the corresponding tweets a month later, when COVID-19 had also become a topic of domestic fear in Western countries. Strong positive sentiment on Weibo would often translate into worship-like appreciation towards characters deemed heroic, like Zhong Nanshan, a Chinese pulmonologist who compiled a COVID-19 diagnosis and treatment protocol during the early stages of the outbreak:

'Mentioning the word hero, I immediately think of Zhong Nanshan. His stalwart image in our hearts cannot be forgotten[1]' (04 February 2020).

Negative sentiment on Weibo would most often express deep hatred for the epidemic and the coronavirus: 'The coronavirus is really abominable, hateful and also fearsome' (5 February 2020).

While posts like those just presented delimited rights and wrongs in the pandemic, overall, COVID-19 content on Weibo focused on a small number of recurring themes. These often formed around the slogans used in the epidemic prevention directives enforced by the Chinese government, for instance: 'million people united as one man' (万众一心) or 'our wills unite like a fortress' (众志成城). The military metaphors typical of the same directives also emerged as a recurring element:

As of February 5, 34 public officials *have sacrificed themselves* in the *fight* against the COVID-19 epidemic. … The current situation of the epidemic is still severe, which requires the party members and all the public to perform their duties and obey the government management. I also hope that all of you will protect yourselves well in this *battle*. (*People's Daily*, 5 February 2020, emphasis added)

While on Weibo COVID-19 content polarised on a small number of themes foregrounded by the Chinese government's epidemic response, on English-language Twitter the discussion was much 'messier'. As a platform used across most of the West, Twitter hosted content on events and response measures taking

Table 4.1: Top gatekeepers of COVID-19 content on Weibo and Twitter during our observation

Rank	Weibo	Twitter
1	@People's Daily	@OH_mes2
2	@Liyongle Teacher	@tedlieu
3	@Global Time News	@Username1
4	@I am Jerry Guo	@Username2
5	@Headlines News	@Surgeon_General
6	@CCTV	@realDonaldTrump
7	@CCTV news	@charliekirk11
8	@Wang Bingru	@Education4Libs
9	@CCTV.Net	@RealJamesWoods
10	@Xinhua Viewpoint	@Username3

Note: For ethical purposes, the handles of all Weibo and Twitter accounts not directly associated to organizations or public figures have been replaced with 'UsernameX'.

place in different countries and regions, for instance via reports released by different international and local organisations and news agencies, declarations made by leading public figures, or comments left by citizens experiencing the start of the pandemic in their everyday life.

On both microblogging platforms, a number of accounts gained high prominence as top gatekeepers, namely as most frequent sources of information, in reposting, retweeting, quoting or mentioning practices (see Table 4.1). On Weibo, these were primarily the official accounts of Chinese official media organisations, such as @People's Daily, @CCTV, @Global Times; @Headlines and @Xinhua Viewpoint[2], which have a close relationship with the Chinese government. For instance, the editor-in-chief and the president of the *People's Daily* are government officials at the ministerial level in China, that is, they cover roles comparable to those of province governors. This political context expects that media organisations, including their social media accounts, align with governmental decision making, which in these early days of the pandemic translated into publicizing the public health measures put in place as part of the country's pandemic response. A small number of online celebrities (for example, 'I am Jerry Guo' and 'Liyongle Teacher') also became top quoted users, with most of their posts forwarding content yet again produced by Chinese official media accounts.

On Twitter, gatekeeping dynamics took a different turn. While, unsurprisingly, the account of former US president Donald Trump made it to the list of top quoted sources, along with those of a few other prominent US figures (for example, democratic congressman Ted Lieu and actor James Woods), a variety of other users also gained prominence as key informers. For instance, @OH_mes2, a news aggregator account mainly sharing reports from South Korean news outlets, became the top gatekeeper of our two-day observation period. These dynamics generated a conversational network that, among other things, covered different areas of the political spectrum (for example, supporting or criticizing Trump's views) and formed around information based on different types of evidence (for example, scientific, anecdotal, politicized).

Platforms, values, and voices

At the outbreak of the COVID-19 pandemic in China and then in the West, Weibo and Twitter played very different roles as participatory platforms embedded in different political economies and sociocultural contexts. On Weibo, centralizing mechanisms concentrated COVID-19 semantics around official, state-controlled sources. This, of course, had a tangible impact on what would become plausible 'COVID-19 information' within the Chinese platform ecosystem and in the broader society. On the other hand, on Twitter influence was primarily exerted by elite and non-elite accounts sharing information that would not necessarily be in line with public health policies. In fact, Twitter-the-corporation soon found itself grappling with mechanisms, indicators, and criteria to identify and signal misleading information circulating on the platform (Twitter, 2020). This was the case, for instance, with an October 2020 tweet by former US president Donald Trump that ended up being flagged as 'spreading misleading and potentially harmful information related to COVID-19' (see BBC News, 2020b).

In their ubiquity, mainstream digital platforms around the world are playing a strong role in the articulation of COVID-19, and have a clear impact on the public understanding of it. However, they do so as part and parcel of existing political economic structures. As is emerging in our observation, Chinese platforms, for instance, while owned by private companies, have close relationships with national regulatory and policymaking authorities. These relationships result in a platform ecosystem that is shaped by the characteristics of state-centred media regulations (Plantin and De Seta, 2019), and that most likely have influenced the public understanding of COVID-19 as a *unifying battle*, while overshadowing other aspects of the pandemic. On the contrary, Twitter's libertarian nature – and overall detachment from state-centred regulatory systems – has facilitated the shaping of COVID-19 as an *infodemic* (Zarocostas, 2020).

The COVID-19 crisis should then prompt us to reflect and act upon the way platform' values – whether state or free market – play a key role in shaping our understanding of public health. We also need to learn more about how users – as humans – carve their voice with or against these values. Both Guo Jing's diary and our two-day journey in pandemic Weibo and Twitter suggest that humans adapt, tinker, resist or subside to the power dynamics enhanced by the structures of governance and the technological affordances emerging in contemporary 'platform societies'.

Notes

[1]   All Weibo content reported in this chapter was translated and paraphrased by the author Zheng Yang.

[2]   The original handles of the top 10 reposted Weibo accounts are: @People's Daily (@人民日报); @Liyongle Teacher (@李永乐老师); @Global Times (@环球新闻); @I am Jerry Guo (@我是郭杰瑞); @Headlines (头条新闻); @CCTV (@中央电视台); @CCTV news (@央视新闻); @CCTV.Net (@央视网); @Xinhua Viewpoint (@新华视点); @IFENG.com (@凤凰网).

References

BBC News (2020a) 'Coronavirus Wuhan diary: living alone in a city gone quiet', *BBC News*, 30 January, https://www.bbc.co.uk/news/world-asia-china-51276656

BBC News (2020b) 'Trump COVID-19 post deleted by Facebook and hidden by Twitter', *BBC News*, 6 October, https://www.bbc.co.uk/news/technology-54440662

Chan, A. (2013) *Networking Peripheries: Technological Futures and the Myth of Digital Universalism*, Cambridge, MA: MIT Press.

Plantin, J.C., and De Seta, G. (2019) 'WeChat as infrastructure: the techno-nationalist shaping of Chinese digital platforms', *Chinese Journal of Communication*, 12(3): 257–73.

Twitter (2020) 'Coronavirus: staying safe and informed on Twitter', https://blog.twitter.com/en_us/topics/company/2020/covid-19.html#misleadinginformationupdate

Willems, W. (2020) 'Beyond platform-centrism and digital universalism: the relational affordances of mobile social media publics', *Information, Communication and Society*, 24(107): 1–17.

Yang, G. (2020) 'The digital radicals of Wuhan', *Center on Digital Culture and Society,* 3 February, https://www.asc.upenn.edu/research/centers/center-on-digital-culture-and-society/the-digital-radical/the-digital-radicals-of-wuhan

Zarocostas, J. (2020) 'How to fight an infodemic', *The Lancet*, 395(10225): 676.

FIVE

Managing Pandemic Risk in an Interconnected World: What Planning a Wedding Shows about Early Responses to the COVID-19 Outbreak

Carlos Cuevas-Garcia

Introduction

Over the past couple of years, a growing number of studies have investigated the social responses and environmental, social, psychological and economic impacts of the COVID-19 pandemic (Jasanoff et al, 2021; Harambam, 2020). As with other contributions to this volume, these studies suggest that the current planetary crisis will generate profound transformations in paid and unpaid labour, mobility, energy use, democratic government, law enforcement, and trust on media and expertise. Some commentators even argue that COVID-19 involves an important reshaping of contemporary globalization, characterized by 'intensifying dynamics of instability, disintegration, insecurity, dislocation, relativism, inequality, and degradation' (Steger and James, 2020: 188). While large-scale and long-term implications of COVID-19 have been explored, the literature has paid less attention to the early days of the pandemic outbreak, when these important and profound transformations were mostly out of sight and had only started to manifest.

Intending to make a modest contribution to fill this gap in the literature, this chapter focuses on the days that followed from 11 March 2020, when the World Health Organization (WHO) characterized COVID-19 as a pandemic. The research method and the data presented here are rather unusual. The account I provide draws from my own experience as the host of an international wedding scheduled for the 21 March 2020 in my hometown, Pachuca, Mexico. Although unexpectedly and unintendedly, organizing an international wedding turned out to be an exceptional way of documenting early responses to the nascent pandemic of a very diverse group of people. Since my partner and I are a German–Mexican couple that lived and studied in the UK for about six years before moving to Germany, our guest list included around 40 people travelling from nearly a dozen countries. Since we were in constant communication with people exposed to different news, local concerns, regulations, health policies and travel restrictions, we were able to get a sense of the rapidly transforming attitudes from individuals and governments in different corners of the planet towards the spread of the virus.

The account I provide here, however, cannot be taken as that of a neutral and innocent observer. The ways in which my partner and I communicated with our guests contributed to some extent to how they perceived and exposed themselves to the risks. In this sense, even in its early days, the pandemic raised important ethical questions about how we should live and act in an interconnected and rapidly changing world. Being the hosts of a large social event expected to take place exactly during the outbreak of a planetary pandemic exposed us to a series of moral dilemmas: should we go on with our plans or should we modify or cancel them altogether? Should we travel to Mexico to discuss with the venue owners and service providers about alternative solutions intending to avoid economic losses, or should we rather avoid travelling and putting others and ourselves at risk? What should we do with the guests who are already in Mexico? Would we be letting them down if we called off the event after all the efforts they made to get there? Our attempts to address these questions depended significantly on what different national governments would do, *and when*, and on our prior assumptions about those political decisions. Thus, the ethical dilemmas we encountered revealed how a private decision – to throw a wedding party or not – was deeply affected by tensions between the global spread of the virus and local ways of dealing with it.

Disrupting normality

In the first two weeks of March 2020, international labour and travel operated as seamlessly and uninterrupted as usual in most of the world's democracies. The movement of people, information, and things was facilitated by successfully integrated economies, infrastructures, and standardized forms of doing things (Barry, 2006). I myself had travelled to Malaga, Copenhagen and Paris during those two weeks. However, on 11 March the WHO made the tragic announcement that COVID-19 had become a global emergency, and next morning US President Donald Trump announced a travel ban for flights coming from Europe – with the exception of the UK – to the US. As we were listening to the news in our apartment in Munich, we received the first cancellation from one of our guests from Germany, whose itinerary included one stop in the US. My partner and I had a flight to Mexico scheduled the next day, so during this time we visited the airline, news, and governments' websites more often than usual. We worried that our flights or those of other guests would face cancellations, were other national leaders to follow Trump's example. However, national governments made decisions differently and taking their time, revealing often neglected differences in forms of reason and sense-making (Jasanoff et al, 2021).

Besides decisions at the national level, the possibilities to travel to Mexico of our guests depended on the measures that their employers implemented. On 12 March, some of our British guests shared with us that their organizations had banned work-related international travel. However, they confirmed that they were still planning to travel to Mexico. The case was not the same for some of our guests from Canada, Peru and the US, whose employers discouraged them from travelling. By that time, it was hard to get a sense of the magnitude of the crisis that was unfolding. Mexico seemed to be a remote and safe place. When the WHO announced that COVID-19 had become a pandemic, Mexico had only 11 *registered* cases, in comparison to the 1,565 in Germany and 456 in the UK. Against that background, the general reasoning among our guests and us was that traveling to Mexico was probably safer than staying in Europe, at least for as long as Mexico was not transitioning into a more advanced phase of the pandemic. Furthermore, the general understanding amongst our guests was that the virus was serious only for people above 60 years old.

Fragmented concerns and understandings

We arrived in Mexico for our wedding during the weekend of 13–15 March, just as the government stated that while they would introduce some new measures a week later, including closing schools and reducing opening times of commercial establishments, they would not close the national borders. Museums and archaeological zones would stay open. Yet over the weekend, our guests' concerns continued to increase, and by Monday, nearly half of our overseas guests had cancelled their travel plans. Their thoughts went rapidly from travelling but shortening their trips to not travelling at all. Some of them feared that their preexisting health conditions would make them more vulnerable to the virus. Others had experienced minor but still concerning symptoms such as coughing and having a sore throat. Some of them worried about having to quarantine on their return home and face troubles at work. Finally, there were some guests changing jobs across countries who had not yet received their residence permits. They worried about not being able to get back into their new countries of residence, and one of them even feared being deported back to Australia.

Alongside these international concerns, some Mexican guests also decided to cancel their attendance to our wedding because of their advanced age and because they considered the government's measures to be too soft and inadequate. Some of them expected more drastic measures, as was the case of Peru, where the government declared a state of emergency on the evening of the 15 March, 10 days after the first case of COVID-19 in that country was confirmed and after only 71 cases had been diagnosed. Our Peruvian friends explained how the citizens feared and anticipated a premature collapse of the healthcare system, which to them justified the decisions that the government had taken, at a time when the virus had not spread too much into their country.

Later that day, the situation deteriorated further, as the German Chancellor Angela Merkel announced severe new measures to control the spread of COVID-19. These included bans on leisure travelling, the restriction of economic activities, and encouraging residents of Germany who were overseas to return home immediately. Her announcement made clear that our wedding celebration plans could not continue, and that we should try to fly back to Europe much earlier than expected.

However, returning turned out to be much harder than we imagined. Besides the mismatching and rapidly changing national policies, commercial airlines' lack of preparedness exacerbated travel difficulties. Numerous flights were cancelled and the airlines were not able to provide clear solutions to their clients. What is more, some companies were not transparent about flight cancellations and neither were they responding to their landlines. Concerned travellers were forced to go directly to the airport to try to secure seats on the few flights that were still on schedule during the following weeks. Some airlines were reassigning people to flights that could be – and as we experienced, were – cancelled 24 hours later, and some considered it reasonable to have waiting lists of more than 50 people per flight, who were forced to go to the congested airport to try their luck and put themselves at the mercy of the airline's lottery system. The German government offered repatriation flights that citizens were obliged to pay for, yet it was difficult to be assigned to these flights. As we have learned from our guests, it took the government about one year to communicate what the compensation costs were.

The ethics of movement in an uncertain world

In only a few days, the human condition of living in an apparently highly interconnected world transformed into a condition of uncertainty, rapid fragmentation and slow reconnection. The story of our called-off celebration points to a number of conclusions about early responses to the pandemic. First, that the globally interconnected world – by systems of travel and communication, a cosmopolitan culture and international standards – is also a tightly-coupled system and one prone to cascading failure, as Perrow (1999) would put it. The pandemic revealed often neglected couplings, for example health and safety restrictions in one place affected travel plans in other places, and people's residence status further complicated travel restrictions. Second, that breakdown was not confined to governments and the public sector, since the private sector also provided slow opaque responses, as was the case of multiple airlines.

Third, we learn that critique to government responses appeared very early and both in cases where responses were judged as too rapid or too slow, too soft or too strong, pointing to several cases of 'civic dislocation' (Jasanoff, 1997): a mismatch between what governments do and what they are expected to do. Fourth, civil dislocation depended on what people saw was happening (or not yet happening) in other places. For instance, after some of our guests saw the responses of the US and Peru, they wondered why Mexico never closed its borders and strengthened its measures. These mismatches further complicated ethical decisions regarding travelling and meeting people despite it being officially allowed, or better avoiding it. In our case, we were lucky to cancel the event five days in advance. When the local government established a ban on all large social events, our guests were already travelling back home to start the long process of learning to live under new conditions.

References

Barry, A. (2006) 'Technological zones', *European Journal of Social Theory* 9(2): 239–53.

Harambam, J. (2020) 'The Corona truth wars', *Science and Technology Studies*, 33(4): 60–7.

Jasanoff, S. (1997) 'Civilization and madness: the great BSE scare of 1996', *Public Understanding of Science* 6(3): 221–32.

Jasanoff, S., Hilgartner, S., Hurlbut, B., Özgöde, O. and Rayzberg, M. (2021) *Comparative COVID-19 Response: Crisis, Knowledge, Politics*, Cambridge, MA: Harvard Kennedy School.

Perrow, C. (1999) *Normal Accidents: Living with High Risk Technologies*, Princeton, NJ: Princeton University Press.

Steger, M. and James, P. (2020) 'Disjunctive globalization in the era of the great unsettling', *Theory, Culture and Society* 3(7–8): 187–203.

PART II

Marginalized Humans

How are different kinds of human included or excluded by the contemporary moment? A range of policies, science advice and services during COVID-19 have had a major impact on the lives of many marginalized groups. Paying particular attention to disabled people, this section will explore the processes of marginalization, how different forms of exclusion and inequality intersect and how these processes are being resisted and transformed. By describing and analysing the lived experience of marginalized people these contributions will provide novel ways of thinking about the human that open up new opportunities for social solidarity and imagining a different future.

SIX

Imperilled Humanities: Locked Down, Locked In and Lockdown Politics During the Pandemic

Dan Goodley and Katherine Runswick-Cole

Introduction

Our contemporary times are marked by human precarity. This precarity is, however, neither shared, universal nor new. We know that Black, disabled and poor people are disproportionately affected by COVID-19. While it is easy to explain this in terms of the unprecedented impacts of COVID-19, we know that Black, disabled and poor people have been disproportionately affected by the pandemic because of years of systemic racism, toxic capitalism and austere underfunding of health, education and social care. These seemingly disposable lives are routinely devalued. How can researchers engage with people who have been further dehumanized by the pandemic? We offer a response to this question in the context of the Black Lives Matters protests of spring 2020, with reference to the global pandemic that has had disproportionate and inequitable impacts, with a specific consideration of the particularly precarious position of disabled, poor and Black people.

Locked down

There is always a danger of reading these tumultuous times as marking a new ground zero of human suffering. Here we are, so a popular narrative tells us, enduring one of the lowest points in human history. But some of us have been here before; times where certain human lives are deemed more worthy than others. Spring 2020. CNN news is playing in the background. We are struck by the public spectacle of the Black Lives Matter protests in the States. It jars with our own solitary reality. Like many around the world, We're staying-at-home, although stuck-at-home feels like a more authentic description. The films rolling before us indicate that the protestors are masked. Social distancing is seemingly being observed. Organized marches wind their ways down major roads in cities across the US. The protestors are clearly putting themselves at risk of transmission. This is not a criticism. It is a commentary on the insecure positions that the activists are prepared to subject themselves to in the service of collectively challenging systemic racism. They seem to be saying: "At risk of the virus? We've been at risk of racism for far longer".

While we are taken by the strength in numbers of the Black Lives Matter activists, the ubiquity of the 'Fuck Trump' banners and the righteous fury on their faces, these displays of dissent contrast so markedly with the current isolation of a number of disabled friends of mine. Many of these mates, we know, are firmly aligned with the Black Lives Matter protests. We can read this in their tweets, Facebook updates and blogs. But we know too that disabled people – especially disabled, poor and Black people – are undergoing a particular kind of pandemic dislocation; unable to take to the streets to support these protests and stuck in domestic contexts that are in themselves alienating and debilitating.

Locked in

To put some context to this discussion it is worth noting that disabled people account for over one billion of the world's population (World Bank, 2011). This is a major minority group. And we know that the vast majority of disabled people live in the majority world, many are people of colour, and a great proportion live in poverty. Attempts to understand disability have at times created a kind of conceptual and empirical whitewashing; ignoring the common ways in which experiences of disability, Blackness and poverty merge together (Miles et al, 2017). While not all disabled people are Black or poor it has become increasingly apparent that when one speaks of disability then this discussion will inevitably engage with Blackness and poverty (Erevelles, 2012).

The primary global response to the pandemic has been a call to shelter-in-place, self-isolate and stay-at-home. Urban areas account for 90% of reported cases (United Nations, 2020). About 60% of the global population is expected to live in cities by 2030, up from 56% today, despite the coronavirus pandemic pushing many who can to look for new homes outside congested urban centres (United Nations, 2020). Overcrowding, rather than density, has enabled the virus to spread, with high population density actually easing the delivery of key healthcare services. Evidence indicates that tackling COVID-19 is more difficult in urban areas due to poverty, poor infrastructure and inadequate housing (United Nations, 2020). The public displays of urban streets-based activism associated with the Black Lives Matter protests capture the centrality of our urban centres, but also contrast with realities of lockdown and with disabled people's experiences of being locked in. From the beginnings of the lockdown in the UK, disabled people with underlying health conditions were advised to shield; which effectively meant staying at home. The home quickly became reimagined as a safe haven. This idealization of the home ignores the fact that many disabled people (who were able to stay-at-home) did so in poor-quality housing; exposed to cold, damp and other hazardous conditions (Clair, 2020). Housing conditions are typically poorest for Britain's 5.5 million private rented sector households. And in 2019, one-quarter of disabled people lived in rented social housing, compared with just 8.2% of the general population (ONS, 2021). Locked in, many disabled people not only had to deal with living in inadequate housing, but also had to cope with the growing existential and material reality that their lives were disproportionately at risk. In February 2021 it was reported that six out of ten people who have died from COVID-19 in the UK are disabled (ONS, 2021); many disabled people from BAME groups and those with learning disabilities/autism being disproportionately impacted (Brothers, 2020). In addition to the loss of life, the pandemic heightened anxiety, isolation, loneliness and mental health difficulties (Mental Health Foundation, 2020). These psychosocial impacts were hugely influenced by the retraction of key health and social care services due to COVID-19; services that had already been decimated by a decade of austerity. There is emerging empirical evidence to suggest that disabled people have become increasingly reliant on their family and other informal carers, driven by the closure or suspension of day centres, day services and large sections of the social care system (Shakespeare et al, 2021). Absent services. Minimal care packages. Absent in/formal carers and loved ones. A lack of human contact due to social distancing. The very real fear of death.

And yet, time and time again, we read on social media the affiliation, support and comradery that many disabled people express in relation to the Black Lives Matters protests.

Blackness, argues Sylvia Wynter (2006:14), tends to be conceptualized as a referent category of the Human Other – an 'unbearable wrongness of being' (2006: 114) – the direct opposite of contemporary interests of Western, White, Bourgeois Man. Blackness, she argues, drawing on Fanon, is experienced as *désêtre* (that is, dys-being); the wrongness of being: the opposite of normative human ontology (where dys is ill, abnormal and bad). In this sense, then, it is hardly surprising to find alliances between Black, poor and disabled people in terms of their politics. Individuals associated with these groups experience *désêtre*. Yet, there is also something distinctive about the impositions placed upon disabled people – especially those who are also poor and Black – as they grapple with the realities of being locked down and locked in.

Lockdown politics

Writing before the pandemic Paul Gilroy (2018: 20) articulates a case for a human project that involves engaging in 'the ongoing work of salvaging imperilled humanity from the mounting wreckage' (2018: 20). This project is needed now more than ever. We need to consider why it is still necessary to assert that Black Lives do Matter, and subject to critical scrutiny the dominant imaginaries and symbolic orders of our societies that devalue Black Life. We need to insert into public discourse and debate the empirical evidence that appears to suggest disabled, poor and Black lives are rendered more expendable and disposable than others. We need to ensure that any intellectual, political or policy encounter with the impacts of COVID-19 attends, from the outset, to questions of disability, Blackness and poverty. We need to find alliances between different political movements and campaigns, and also acknowledge disparities and differences. We need to acknowledge that politics takes place in the street and in the home.

One conceptual approach that might inform the ongoing work of salvaging imperilled humanity from the mounting wreckage lies in sociogeny. This perspective, developed by Sylvia Wynter (2006) in response to Frantz Fanon (1993), demands that we unpack the social, historical and cultural constitution of humanness. In counter-distinction to phylogeny (the study of evolution of the species) and ontogeny (the biological development of an individual organism), Wynter's sociogeny unpacks the social, historical and cultural constitution of humanness through which disability, Blackness and poverty become to be known and lived. Reading disability,

Blackness or poverty through the methodologies of phylogeny or ontogeny will reduce these to accounts of evolution or biology. Read through sociogeny we come to appreciate the ideological, material and symbolic constitution of Black, disabled and poor lives. Sociogeny recognizes that there is no experience of human biology that is detached from sociology; 'our ontological experience of humanness is inherently shaped by our subjective experience of social relations in the world' (Dy and Jayawarna, 2020: 392). A sociogeny of COVID-19 would not only understand this phenomenon as a fundamentally social phenomenon, but would also recognize that some humans are already imperilled before they even come into contact with this disease; whether this be on the street or in the home.

References

Brothers, E. (2020) 'Building back better: disabled people and COVID-19', *Community,* 3 December, https://community-tu.org/building-back-better-disabled-people-and-covid-19/#a10287c4

Clair, A. (2020) 'Homes, health, and COVID-19: how poor housing adds to the hardship of the coronavirus crisis', *Social Market Foundation,* https://www.smf.co.uk/commentary_podcasts/homes-health-and-covid-19-how-poor-housing-adds-to-the-hardship-of-the-coronavirus-crisis/?doing_wp_cron=1586254540.1653978824615478515625#_edn13

Dy, A. and Jayawarna, D. (2020) 'Bios, Mythoi and women entrepreneurs: a Wynterian analysis of the intersectional impacts of the COVID-19 pandemic on self-employed women and women-owned businesses', *International Small Business Journal: Researching Entrepreneurship,* 38(5): 391–403.

Erevelles, N. (2012) *Disability and Difference in Global Contexts: Towards a Transformative Body Politic,* London: Palgrave.

Fanon, F. (1993) *Black Skins, White Masks,* 3rd edn, London: Pluto Press.

Gilroy, P. (2018) '"Where every breeze speaks of courage and liberty": offshore humanism and marine xenology, or, racism and the problem of critique at sea level', *Antipode,* 5(0): 3–22.

Mental Health Foundation (2020) *Coronavirus: The divergence of mental health experiences during the pandemic*, London: Mental Health Foundation.

Miles, A., Nishida, A. and Forber-Pratt, A. (2017) 'An open letter to White disability studies and ableist institutions of higher education', *Disability Studies Quarterly,* 3(3), http://dsq-sds.org/article/view/5997

ONS (Office for National Statistics) (2021) 'Updated estimates of coronavirus (COVID-19) related deaths by disability status, England: 24 January to 20 November 2020', https://www.ons.gov.uk/peoplepopulationandcommunity/birthsdeathsandmarriages/deaths/articles/coronaviruscovid19relateddeathsbydisabilitystatusenglandandwales/24januaryto20november2020

Shakespeare, T., Watson, N., Brunner, R., Cullingworth, J., Hameed, S., Scherer, N., Pearson, C. and Reichenberger, V. (2021) Disabled people in Britain and the impact of the COVID-19 pandemic, *Preprints 2021*, 2021010563.

United Nations (2020) Policy brief: COVID-19 in an urban world, 28 July, https://unsdg.un.org/resources/policy-brief-covid-19-urban-world

World Bank (2011) *World Report on Disability*. Geneva: WHO.

Wynter, S. (2006) 'On how we mistook the map for the territory, and reimprisoned ourselves in our unbearable wrongness of being, of *Desêtre*: Black studies towards the human project', in L.R. Gordon and J.A. Gordon (eds), *Not Only The Master's Tools: African-American Studies in Theory and Practice,* Boulder, CO: Paradigm Publishers, pp 107–69.

SEVEN

"Why Would I Go to Hospital if It's Not Going to Try and Save Me?": Disabled Young People's Experiences of the COVID-19 Crisis

Kirsty Liddiard, Katherine Runswick-Cole, Dan Goodley
and the Co-Researcher Collective: Ruth Spurr,
Sally Whitney, Emma Vogelmann, Lucy Watts MBE
and Katy Evans

Introduction

Throughout our co-produced project, *Life, Death, Disability and the Human: Living Life to the Fullest* (ESRC 2017–2020), disabled children and young people living with shortened life expectancies have readily emphasized their human worth, value, and desire for the future. They have done so in disabling cultures that routinely deny them opportunity, access, and expectation. Perhaps not surprisingly, our conversations with disabled young people – and our interpretation of them – became more complex upon the onset of the COVID-19 global pandemic. Suddenly thrown into a moment where all lives became (more) vulnerable – an already-lived reality of many of the young people in our project – it was also a time where cultures of ableism and disablism were made more explicit, and existing inequalities exacerbated. For clarity, we use the terms 'ableism' and 'disablism' throughout this chapter. Ableism relates to the material, cultural and political privileging of ability, sanity, rationality, physicality and cognition (Braidotti, 2013), while disablism is the resultant oppressive treatment of disabled people (Slater and Liddiard, 2017).

In this chapter we share co-researchers' own blog posts and writings on their experiences of *living* through a pandemic. Importantly, young people's voices explore the (new) ways in which they have made sense of risk and threat, from the virus itself, but also from discriminatory emergency policymaking, compromised access to health resources, and a general lack of governmental support – all of which has affirmed the disposability of disabled and vulnerable lives in contexts of dis/ableism.

"I know full well in this COVID-19 pandemic that my life is not one that will be saved": managing discourses of human worth

COVID-19 began with early public health messages that only the elderly and those with existing health conditions are most at risk of serious illness or death. Such ontologically violent messages quickly sought to reassure an overwhelmingly anxious public at the expense and distress of some of those considered the most vulnerable. Further, government ministers affirmed herd immunity as an initial key strategy: the concept of allowing publics to be exposed to a virus, in the hope that spreading it among those who are at low risk means that a large part of the population becomes immune. Such a Darwinist approach was early affirmation for disability communities of the extent to which disabled and chronically ill lives are routinely devalued in dis/ableist cultures. Of course, the pandemic has globally engendered feelings of fragility – COVID-19 will undoubtedly lead to a host of physical and mental health anxieties for many people. But for disabled people this affective positioning has been further shaped by a dominant cultural imaginary in which their very human worth has been called into question (Goodley et al, forthcoming).

Ideas of herd immunity, however, quickly subsided in the UK when it was strongly refuted by the World Health Organization (WHO), and deemed risky and unscientific by Imperial College epidemiologists and virologists. Yet for the disabled young people in our project such public discourses cemented a very sudden *new* precarity. This was quickly furthered by seemingly routine discussions over a lack of resources in the Prime Minister's COVID-19 daily briefings and in the media – the numbers of critical care beds; access to ventilators; and the seemingly rational decision making around the distribution of resources. Understandably, such troubling conversations had deep affective meaning for our disabled co-researchers, and a key question surfaced quickly within the Research Team: which of us will get life-saving treatment and which of us will not?

As the project's Lead Co-Researcher, Lucy wrote in her blog early on in the pandemic (Watts, 2020: np):

> I have had to make the difficult step to say that if I get COVID-19, I will not be going to hospital. Why would I go to hospital if it's not going to try and save me? … they won't ventilate me anyway … there's no point in me going to hospital. … I have to accept my life will not be saved.

Furthering this worry, new NICE guidelines (announced suddenly on Friday 20 March 2020; NG 159) brought together existing national and international guidance and policies and advice from specialists working in the NHS from across the UK. NICE guidelines stated that on admission to hospital, medical professionals will assess all adults for frailty (irrespective of age and COVID-19 status); consider comorbidities and underlying health conditions; and use the Clinical Frailty Scale (CFS) – originally developed for dementia – for frailty assessment. Despite being a deeply troubling measurement of human worth, NICE placed the CFS at the centre of admission to critical care in the first wave of the pandemic. In short, a score of five or below makes you more likely to access critical care, while a score of six or above makes this less likely. Lucy stated in her blog the affective realities for her as someone living with life-limiting and life-threatening impairments (Watts, 2020: np.):

'On the clinical frailty scale, I'm a 7 or even 8. They don't save people at that level, when forced to choose between them and a person with less needs and a higher survival rate.'

Lucy's words here reveal the emotional and psychic impacts of dis/ableist discourses of human value during a public health crisis. Such guidelines make visible the inherent ableism that determines whether someone lives or dies based on the support they require in everyday life. For example, as a Research Team we asked, how did we get to the point where 'frailty' that 'impairs shopping, walking outside alone, meal preparation and housework' (#5 CFS) is integral to human value? And where those who 'often have problems with stairs and need help with bathing and might need minimal assistance with dressing' (#6 CFS) have less of a right to survival than those who are deemed 'robust, active, energetic and motivated' (#1 CFS)?

"How are disabled people expected to protect themselves?": making sense of the new risks of care and caring

Co-Researchers also drew our attention to the difficult practicalities of managing care during a global pandemic. Co-Researcher Emma outlined the paradox COVID-19 presents for those with chronic health conditions who are the recipients of personal assistance and care: '… while on the one hand, we're most at risk of becoming seriously ill if we contract COVID-19, we're also often the least able to self-isolate' (Vogelmann, 2020; np). Such practicalities were undergirded by a lack of governmental attention and awareness to the new dangers of care provision and the routine de-prioritization of disabled people, their families and carers. Within weeks of the pandemic beginning, the UK government had ratified the Coronavirus Act (2020), which formally suspended the Care Act 2014, suddenly reducing disabled people's rights to social care. This, combined with the lack of provision of PPE in care homes and for home carers and personal assistants (Shakespeare et al, 2021), increased the very literal dangers of COVID-19 for disabled people and their families, as Emma articulates (Vogelmann, 2020: np):

> I have a team of six regular carers who come in and out of my house every week, who could potentially bring in coronavirus. This causes me a lot of worry, but there is little I can do about it. I have asked my carers to wash their hands when they come in, and to let me know if they or someone they've been in contact with has symptoms ... given this, how are disabled people expected to protect themselves?

Emma later emphasizes how such risks are exacerbated by a lack of clarity from the government, arguing the 'need for NHS guidance for care staff in home [domestic] settings; advice for disabled employers who are also high-risk individuals; and acknowledgement that it is impossible for us to completely self-isolate (Vogelmann, 2020: np).

"What life is really like for us": everyday experiences of isolation

Within disability communities, there has been an affirmative sense of recognizing that disabled people and their families often already possess the skills it takes to survive risk, threat and isolation. For many, as Co-Researcher Ruth says, it is a daily reality: "... with all this talk of self isolation ... the world is finally seeing a little glimpse of what life is really like for us. For the unseen, the chronically unwell, the bed and housebound, the invisible ones not seen for weeks or months at a time".

For Ruth, regular UK lockdowns and other safety measures offer a window into her disability experience – a chance for the sheer labours of staying alive when you live with life-threatening impairments to become recognized and shared by others. Yet for Lucy, COVID-19 once again shored up the need to affirm her very humanness: her (social) value, her contributions, her desires and dreams for the future: '… I do not accept that my life is less worthy in the eyes of who should be saved … I have so much to do, so many plans, so many ideas. My life matters, my life is worthy, my life is valuable, my life is appreciated' (Watts, 2020: np.)

Lucy's words affirm a key finding from our project – that routinely stressing the value of your own life, self and future, is a form of emotional labour routinely carried out by disabled young people in contexts of dis/ableism (Whitney et al, 2019). For our Research Team, such labours have been significantly exacerbated by COVID-19 and governmental responses to it. Importantly, through their own writing, our Co-Researchers are engaging in what Shaw (2012: 42) calls discursive activism: 'speech or texts that seek to challenge opposing discourses by exposing power relations within these discourses, denaturalising what appears natural'.

Conclusion

As the Co-Researchers' testimonies in this chapter have shown, the pandemic context once again means that disabled people, our allies, and communities need to take urgent action to humanize ourselves and our lives, particularly with and for those deemed the most vulnerable among us. For Lucy, we must emphasize that we are valuable and valued human beings; that being vulnerable does not equate with being less than human. For Emma, that we have meaningful lives worth protecting and saving. For Ruth, that there is little cultural understanding of the often isolative realities of disabled lives. That, ultimately, ableism – both inside and outside the precarious context of crisis – readily dehumanizes and demarcates our lives as disposable.

References

Braidotti, R. (2013) *The Posthuman*, Cambridge: Polity Press.

Goodley, D. et al (forthcoming) 'Affect, dis/ability and the pandemic', Special Issue: New Dialogues between Medical Sociology and Disability Studies, *Sociology of Health and Illness.*

Liddiard, K. and Slater, J. (2017) ' "Like, pissing yourself is not a particularly attractive quality, let's be honest": learning to contain through youth, adulthood, disability and sexuality', Special Issue: Disability and Sexual Corporeality, *Sexualities* 21(3): 319–33.

NICE (National Institute for Health and Care Excellence) (2020) 'COVID-19 rapid guidelines: critical care', https://www.nice.org.uk/guidance/NG159

Shakespeare, T., Watson, N., Brunner, R., Cullingworth, J., Hameed, S., Scherer, N., Pearson, C. and Reichenberger, V. (2021) 'Disabled people in Britain and the impact of the COVID-19 pandemic', *Preprints 2021*, 2021010563.

Shaw, F. (2012) 'The politics of blogs: theories of discursive activism online', *Media International Australia*, 142(1): 41–9.

Vogelmann, E. (2020) 'How are people who need carers supposed to self-isolate?', https://www.independent.co.uk/voices/coronavirus-self-isolation-nhs-advice-disabled-carers-a9424431.html

Watts, L. (2020) 'These are our realities, and we are the casualties – COVID-19', https://www.lucy-watts.co.uk/2020/03/26/covid-19-blog/

Whitney, S., Liddiard, K., Goodley, D., Runswick-Cole, K., Vogelmann, E., Evans, K., Watts (MBE), L. and Aimes, C. (2019) 'Working the edges of posthuman disability studies: theorising with young disabled people with life-limiting impairments', *Sociology of Health and Illness*, 41(8): 1473–87.

EIGHT

Science Advice for COVID-19 and Marginalized Communities in India

Poonam Pandey and Aviram Sharma

Introduction

The relationship between modern science and the public is complex even in 'normal' times (Nowotny et al, 2013). In times of pandemic, when knowledge uncertainties magnify, and decisions need to be taken in extremely pressing situations, this relationship needs extra care, caution and nurturing. In the absence of such care, science advice might do more harm than good, and perhaps will end up creating multiple new vulnerabilities, marginalities and loss. The primary focus of this chapter is to engage with the relationship of science and its public. What kind of public did science-based advice imagine while providing health advisories to prevent and contain a COVID-19 outbreak in India? Who did it include and who got marginalized in the process?

The analysis draws theoretically from Mike Michael's (2009) conception of Public-in-General (PiG) and Public-in-Particular (PiP), which are associated with and emerge in response to different versions of science. While PiGs are ignorant yet cooperative, they fit more closely to the official narratives and emancipatory purposes of science; PiPs are often regarded as interest–driven, emotional and disruptive (Michael, 2009).

Science advice for the pandemic and its 'public'

The science advice to curb the spread of the pandemic in India was designed and disseminated along the lines of the traditional model of science communication. This deficit model relies on the idea that the source of all disagreements between state/science and citizen/public lies in the lack of proper information and knowledge (Sturgis and Allum, 2004). With the 'right' information, the public will behave in desired and predictable ways, reducing chances of misinformation and confusion-led chaos, enabling governmentality. Traditional science communication-led interactions with the public were focused on making the public aware and better informed.

In the COVID-19 case, the public was envisioned as Public-in-General and assumed to be completely ignorant, lacking scientific knowledge and understanding. Public health advisories were disseminated through multiple visual and verbal platforms to promote awareness and bridge the information deficit. Scientific institutions (Indian Council of Medical Research), government departments and ministries (Ministry of Health and Family Welfare, Department of Science and Technology) took the lead at the central level, while other state-based institutions (district level administrative bodies) were roped in to communicating the health advice at the local level. Street signs, billboards and walls of government hospitals, schools and local administrative offices were filled with information displays presented through colourful cartoons and attractive slogans and messages. There were loudspeaker announcements, multiple times a day, where information was transformed into catchy song lyrics and catchphrases to attract public attention.

In addition, IT-based communication technologies/tools/ services (such as SMS, caller tune, WhatsApp, specific apps like Aarogya Setu) were deployed to disseminate the health advisories. The content of the information provided to the public was organized along six major themes: 1) a statement about the ongoing pandemic and its spread; 2) symptoms of

COVID-19 infection; 3) the mechanisms through which the infection spreads; 4) precautionary measures; 5) fines and legal actions on people not following the advice; 6) collective and individual roles and responsibilities of the public in defeating the virus.

Although each aspect of the content of the information provided for preventing the spread of COVID-19 needs discussion, this chapter will focus primarily on the precautionary measures that were a central feature of the science advice to the public. Key aspects of protective measures were staying at home, wearing face masks in public places, and maintaining social distancing, sanitation measures that include covering the face while sneezing and coughing, washing hands every 20 minutes with soap or alcohol-based hand sanitizers, and self-reporting and visiting a nearby healthcare centre, if any of the COVID-19 symptoms are noticeable in family members, co-workers and neighbours. Implicit in these science advice measures is the imagination and construction of a uniform Public-in-General (PiG) that could be ordered through effective information dissemination measures (Michael, 2009). In this characterisation of the public, internal differentiations, based on socioeconomic and cultural settings and spatial inequalities, which hinder action on such advice, are downplayed.

As long as these measures were voluntary, not much discussion emerged in public forums. However, on 22 March 2020, with very little preparation, the government of India declared a three-week lockdown (Abraham, 2020). This lockdown was based on advisories/projections offered by a few reputed international science organisations and the health advisory group (Corona Task Force) created by the Prime Minister's Office to fight the pandemic. The lockdowns were subsequently extended till June 2020 and were enforced using heavy police force, monetary fines and legal actions (Thomas, 2020). The national-level lockdown severely affected the social and economic life of the public. Within this state-enforced practice of science advice aimed at Public-in-General

(PiG) there emerged many Publics-in-Particular (PiPs), that challenged and deconstructed the idea of PiG embedded within the design of science advice.

These PiPs differed in their relationship with science advice in terms of their capacity and capability to follow these measures and their trust of scientific institutions and science advice. Critically, the worldviews and everyday decision making of different PiPs were heavily influenced by multiple other vulnerabilities.

Publics in response to science advice

The obedient subject

This category of public is constituted by the large and growing section of the urban and semi-urban middle class who have a stable income and expandable savings, and are mostly employed in the formal sector of the economy. Middle class is defined using various variables and there is no consensus on its size; according to some projections it is estimated at between 28 to 40 % of the overall population (Aslany, 2019). It seems that the science advice was designed by keeping this population in mind. This group of the middle-class public (dominantly the salaried class) had the capacity to work from home, by making initial investments and adjustments, storing month-long supplies, ordering services online, and participating as obedient citizens of the state in the collective national 'war against the coronavirus'.

The disobedient migrant/labourer

Amid the media spectacles of proactiveness and strong government interventions, one of the most severe migrant crises emerged and was compared to the migration of millions of people after the partition of India and Pakistan (Mukhra et al, 2020). The informal sector contributes around 50% of the overall value in the Indian economy, and informal workers account for around 92.4% in the economy (Murthy, 2019). A large section of the public employed in the informal sector regularly and repeatedly moves between their native place (place of original residence) and urban centres (place of employment). Most of these groups live in unplanned colonies, in squatter settlements, in fringe areas of the cities having inadequate access to proper urban infrastructure. Access to good sanitation and clean water remains a perennial challenge to these groups in most of the cities in India (Sharma and Harvey, 2015). Following social distancing and sanitisation-related advisories in such situations is not a choice but a luxury that people living in these settlements can't afford even in 'normal' times.

Scientific advice that informed the lockdown created a situation of prolonged uncertainty with regard to their status of employment. The health risks due to coronavirus infection seemed smaller than the apparent socioeconomic vulnerabilities which were induced by the lockdown. To overcome this, large sections of this group started returning to their native places. Shelter and food were not perceived as a major and immediate challenge due to the social capital and embeddedness of this populace in the local economy of their native place. As a result, despite reassurance from state agencies for extending support, thousands of men, women and children journeyed on foot for several hundred kilometres and used other means to return to their native places, enduring harsh weather, lack of food and water, and state and police brutalities (Mukhra et al, 2020). Vulnerability due to existing informality, precarious economic situations, and distrust of the state in general, made people in this category defy science advice and emerge as a dis-obedient public.

The critical citizen

A vocal and visible minority, this category of the public could be identified by their constant effort to demand accountability from the state-science complex. Consisting mainly of the educated urban middle class, this public emerged in different forums to critique the inadequacy of the government and scientific institutions in dealing with the challenges posed by the pandemic. The public in this group vocalized their discontents through social media platforms such as Twitter, blogs and vlogs. Apart from public intellectuals, journalists and media influencers, many common people also responded to official channels of the state-science institutions demanding accountability for the migrant crisis, vaccine status, testing inadequacies and lack of protective gear for frontline workers[1]. Unimpressed by the spectacle of proactiveness of the government disseminated through popular media, this public employed their agency as empowered citizens and took to alternative media platforms to engage with the state-science complex, rather than be obedient spectators.

The invisible sufferers

Beyond the diversity of visible public-in-particular, which emerged in response to science advice, there are multiple groups of people who are subjected to the negative impacts of lockdown. This group of PiP suffered the impacts of science advice-led lockdown in silence. They include people with terminal illnesses[2] requiring regular visits to hospitals, children from low socioeconomic backgrounds enrolled in government schools, victims of domestic abuse, tribal and indigenous communities, and small and marginal farmers. Structural inequalities and fault lines were laid open, yet the policy response failed to adequately address the needs of this group of PiP. The halting of public transportation, closing of markets, and shifting of the treatment of physical classes to online platforms disproportionately affected the marginalised communities. For instance, in the absence of resources to buy Android mobile phones and laptops, education became inaccessible to a large majority of such children, many of whom will face hardships in the future; especially girl students, and some might never return to school again.

Conclusion

This chapter demonstrates how different kinds of publics come into being in relation to science advice for COVID-19 in India. The unifying vision of public-in-general that informed top-down science advice and state measures on lockdown largely imagined an ignorant public lacking the 'right' information. As a result, the majority of steps taken by the state-science complex were directed towards building public capacity to access and consume the 'right' information and science advice. This singular vision of the public-in-general neglected and undermined the diversity of social groups in terms of their socioeconomic situation, spatial embeddedness and cultural and historical situatedness in relation to science. Incorporating the needs and particular situations of diverse groups of publics is crucial to consider while making science-based advisory decisions which affect their life and livelihoods. There is a need for the recognition of multiple vulnerabilities that are often excluded in narrow, elitist and top-down science and technology-based managerial approaches. In order to be inclusive and responsive, science advice should rely on learning across disciplinary boundaries and epistemic communities.

References

Abraham, T. (2020) 'COVID-19 communication in India', *Journal of Communication in Healthcare*, 13(1): 10–12.

Aslany, M. (2019) 'The Indian middle class, its size, and urban-rural variations', *Contemporary South Asia*, 27(2): 196–213.

Michael, M. (2009) 'Publics performing publics: of PiGs, PiPs and politics', *Public Understanding of Science*, 18(5): 617–31.

Mukhra, R., Krishan, K. and Kanchan, T. (2020) 'COVID-19 sets off mass migration in India', *Archives of Medical Research*, 51(7): 736–8.

Murthy, S.R. (2019) 'Measuring informal economy in India', Seventh IMF Statistical Forum, Washington, DC.

Nowotny, H., Scott, P.B., and Gibbons, M.T. (2013) *Re-thinking Science: Knowledge and the Public in an Age of Uncertainty*, Hoboken, NJ: John Wiley & Sons.

Sharma, A. and Harvey, M. (2015) 'Divided Delhi: bricolage economies and sustainability crises', in M. Harvey (ed), *Drinking Water: A Socioeconomic Analysis of Historical and Societal Variation*, London: Routledge.

Sturgis, P. and Allum, N. (2004) 'Science in society: re-evaluating the deficit model of public attitudes', *Public Understanding of Science*, 13(1): 55–74.

Thomas, L. (2020) *How India's lockdown has affected mental health*, https://www.news-medical.net/news/20200527/How-Indias-lockdown-has-affected-mental-health.aspx

NINE

Pandemic Satire and Human Hierarchies

Tanya Titchkosky

Q: Why don't chefs find coronavirus jokes funny?
A: They're in bad taste.

Jokes and political satire have not been on lockdown during the COVID-19 pandemic. Books, blogs, memes and one-liners abound, all poking fun at the terrifying and deadly matter of the viral scourge that has killed millions of people around the globe. Disabled and older people, especially those who are poor and/or racialised, have been hardest hit by the pandemic, and have also been made the targets of Western racist humour, as well as anti-racist satire. My particular concern is with the paradox of politically woke satire carrying out its critique through the denigration of disability. Such political satire exposes racist and bogus beliefs circulating in the pandemic, but it does so by using terms that reference, sometimes with vehement hatred, disabled people. This is more than blaming the victim; it fails to acknowledge how social responses to the pandemic have had a devastating effect on disabled people, who account for six out of ten COVID-19 deaths (Aljazeera, 2021). Through an analysis of one example of this common satirical trope, I aim to show that a critical disability studies perspective can awaken political imagination by not reproducing traditional hierarchies of humanness.

Consider this:

Outbreak

TORONTO – REPORT: OUTBREAK OF IDIOCY SPREADING 10000 TIMES FASTER THAN CORONAVIRUS

Public health officials in Toronto have confirmed its first 50,000 cases of being a misinformed fuckwit as xenophobic conspiracy theories and tales of false cures continue to spread across social media.

"Becoming a complete moron during an infectious disease outbreak is far more viral than we first thought," said Dr Jeanne Smith of Toronto Public Health. "Fact resistance is abnormally high especially among the dullard population, and the bottom 5% of your graduating high school class."

Tens of thousands of people were affected by a novel fake news claim that the Chinese government was developing coronavirus at Canada's National Microbiology Lab leaving at least 10,000 people stupider.

Patients are usually asymptomatic until they open their mouths or start tweeting.

Aunts spreading rumours about 100% natural cures for the virus on Facebook have been quarantined while racist uncles at dinner tables were ball-gagged as a precaution.

"Our epidemiologists are working hard to identify idiot zero, but there might be more sporadic outbreaks of coronavirus-related imbecility", added Dr Smith.

Meanwhile, health officials are dreading teaching the population a complicated prevention technique: washing your hands.

Source: Beaverton, 2020

Within 24 hours, this satirical spoof had obtained 159,000 likes and was shared by many, including my politically active left-leaning Facebook friends. Bad taste, these friends eventually agreed. Importantly, the article remains in circulation, and so too does the ruthlessly callous and derogatory circulation of a panoply of old and new conceptions of intellectual differences.

Outbreak can be read as attempting to combat a veritable explosion of racist media renderings of the coronavirus pandemic and health crisis. Recall that racism has been perpetrated not only by fringe groups but by the mainstream media that initially named the virus after a province in China, and regularly attributes the pandemic's genesis to Chinese cultural practices. The spread of racist rhetoric replete with yellow scourge stereotypes of the past finds its further expression in the ongoing violence perpetrated against people of Chinese descent across the West (Chinese Canadian National Council, 2021). It is into this cultural scene that antiracist satire aims to direct its incisive intervention.

Outbreak depicts racist false causes given for the pandemic as resulting from 'coronavirus-related imbecility'. This depiction positions author, reader and anyone who finds this funny as naturally superior to those depicted as intellectually inferior who represent the 'real' contagion threat of racism. Still, who is suggesting that 'misinformed fuckwit', 'complete moron', 'dullard population', the 'bottom 5% of your graduating high school class', and the 'stupider', 'idiot zero', who caught 'coronavirus-related imbecility', are to reference anyone other than racist aunts, uncles, and tens of thousands of others who invoke hatred toward people understood as Chinese as a way to explain the pandemic? Interestingly, it is only the term 'misinformed fuckwit' that hides a little of its long history of medical nomenclature referencing people with intellectual differences and impairments – the 'halfwit' once called the 'idiot'. These archaic medical terms, stripped of any official diagnostic power, imbued with denigration, are now commonly used in everyday life. In *Outbreak,* these terms are used to suggest that racism is stupid, and those who practice it a victim of their own 'imbecility', that is, intellectual limits.

The dynamics of such humour bring out deep political divisions; for example, satire plays with multiple existing inequalities magnified by the pandemic. Play is key here. Satirical play can take what is prevalent and powerful, find a pleasure in critically exposing the workings of power, while perhaps provoking new possibilities for life together. Pandemic satire plays with deep divisions in order to toss them into unexpected and potentially powerful critiques, even as satire skirts the edges of which humans should concern us and which not. Yet, pandemic satire not only addresses but also remakes existing fissures and fractures of society.

Garnering a laugh from many, satire implies a good reason for bad humour since it turns a profit for the publication. Other good reasons for bad humour can also be found; after all, jokes and political satire may be understood as the provision of distance between us and the threat (COVID-19), offering a kind of coping strategy that brings out various sentiments placed between an 'us' who is not ill or bad and a 'them' who is. There are, however, many threats that have come to the fore during the pandemic, not the least of which are ableism and racism as ways of making sense of who is to blame for our current deadly state of affairs. Likely, ableist and racist satire will not cease with the end of the pandemic, since in addressing racism the sense of the lesser human is nonetheless reproduced. So long as hierarchies of humanness are in play, racism remains an ever-present possibility (Gilroy, 2005).

The ideology of the less-than-human in *Outbreak* re-inscribes a restricted human imaginary by suggesting, over and over again, that one is either empowered by the capacity to perceive antiracist truth or is not and, if the latter, one is perceptually impaired because intellectually inferior. Naturalising the perception (or not) of racism, like the colour-blind tropes of the past, replaces one cultural formation of human inferiority with another. In so doing, there may be a pleasurable collective flash of mutual recognition – *we* know racism when we see it and *we* know the fuckwits who perpetuate it – but, then, a sense of *we* solidifies, establishing itself against the 'morons', 'idiots', and any others who catch 'racist-imbecility'. The belief that mental differences and intellectual impairments, treated as a kind of natural inferiority, lie behind racist perception, re-inscribes the superiority of those flexing their 'natural' capacity to perceive racism.

People with intellectual impairments constitute a major group of those whom the virus has hit, and are most likely to face severe illness and death. Moreover, this is the group of people that triage protocols have been most likely to exclude from treatment or to be admitted into hospital with do not resuscitate orders. Catherine Frazee (2009: 122), writing during a previous viral outbreak in Canada, made it clear that being disabled in 'dangerous times' is facing one's possible exclusion from care:

> The criteria specify that patients will be excluded from admission/transfer to Critical Care if any of a number of conditions are present, including the following:
>
> * advanced untreatable neuromuscular disease;
> * severe and irreversible neurologic event/condition;
> * severe cognitive impairment.

Nothing fudgy or generic here. Spelled out quite clearly, in fact. There it is, that same familiar melody, that cold and clammy farewell handshake, utilitarian-style. Sorry, we've done all that we could for you and your kind. But the waters are rising, and there are others in desperate need. There are others that we must save, because they will live long and rich lives (homonym intended). Difficult choices must be made. Surely you understand.

Yes, I believe we do understand. Just keep washing our hands.

The absolute necessity to address racist beliefs and actions remains in this cultural scene, and remains deeply intertwined with disability and impairment in innumerable ways. Why then the political satire trope that functions on the basis of an imagined radical divide between racism and disablism?

Naturalising the antiracist perspective serves a false consciousness regarding human life, since it cannot perceive its own reliance on a conception of '... Man, which overrepresents itself as if it were the human itself' (Wynter, 2003: 260). Ideologies of bio-economic-functional 'Man', as if it were the human itself, must be addressed at every turn if our relations are going to be revolutionised, and this requires awakening the capacity to orient to perception as it is informed by ideology. Political satire can make us pay attention to cultural commitments, such as racism; yet this capacity to perceive is still based on a belief that there are lesser-others but their identity has been mistaken. Once again, some people are regarded as naturally inferior, a belief in hierarchies of humanness is reproduced, and remains at the ready for making sense of threatening events of any kind.

Confronted with the ableist use of disability identities in order to battle against racist narratives of the pandemic's origin and deadly devastation, what is to be done? Is political satire possible without relying on some group of humans as naturally less-than others? Whatever the answers to these questions, it would take a revolutionary form of imagination, an insertion of the truly unexpected and denaturalised notion of human inferiority, to upend the ongoing perpetuation of hierarchical understandings of humanness. Incisive political commentary that doesn't rely on the reproduction of hierarchies of the human might be impossible – but it is a challenge I wish satirists were more alive to. To create a momentary spark of recognition, a pleasure and laughter, not based on the assumption of a lesser human, is undoubtedly difficult. At the same time, critical disability studies offer a counter-difficulty, upending recognition so that a groan, not a laugh, is evoked when disability is made to serve as the scapegoat, when hierarchies of the human are re-etched.

It is ironic that we find ourselves building, again, and again, hierarchies of humanness as a source of pleasure that is, in fact, a trap for our lives together. Still, it seems, from within this trap, what counts as human and His lesser others must be addressed at every turn if our relations are to be revitalised. This requires that we nurture the revolutionary pleasure in perceiving perception as it is informed by cultural ideology. But, for now, we simply have political satire buffering against change by reproducing the degradation of some people as a pleasure for others, and making the reproduction of hierarchies of the human a defining feature of satire itself.

Such satire seems like such a terrible joke.

References

Aljazeera (2021) 'COVID's disabled victims', *Aljazeera,* 4 February, https://www.aljazeera.com/program/people-power/2021/2/4/covids-disabled-victims

Beaverton (2020) 'Outbreak of idiocy spreading 10,000 times faster than coronavirus, Toronto', *Beaverton*, 28 January, https://www.thebeaverton.com/2020/01/report-outbreak-of-idiocy-spreading-10000-times-faster-than-coronavirus/

Chinese Canadian National Council (2021) 'Report', 25 March, https://mcusercontent.com/9fbfd2cf7b2a8256f770fc35c/files/35c9daca-3fd4-46f4-a883-c09b8c12bbca/covidracism_final_report.pdf

Frazee, C. (2009) 'Disability in dangerous times', *Journal of Developmental Disabilities,* 15(3): 118–24, https://oadd.org/wp-content/uploads/2009/01/Frazee_15-3.pdf

Gilroy, P. (2005) *Postcolonial Melancholia,* New York: Columbia University Press.

Wynter, S. (2003) 'Unsettling the coloniality of being/power/truth/freedom: towards the human, after man, its overrepresentation – an argument', *CR: The New Centennial Review*, 3(3): 257–337, https://doi.org/10.1353/ncr.2004.0015

PART III

Biosocial Humans

How is new scientific knowledge, as well as its popularization and application in various biomedical technologies, changing ideas of what it means to be human in the 21st century? These include the reworking of long-established ideas of biological determinism in understanding both health and behaviour. The pandemic is shifting these discourses and the way in which we value the life of different humans, reinforcing existing social divisions and creating new ones.

TEN

Genomic Medicine and
the Remaking of Human Health

Paul Martin

The contemporary development of genomics – the decoding of our DNA – marks an important turning point in how we understand the human. This is nowhere more apparent than in the UK, which is leading the world in investing in and developing powerful genomic technology platforms that may have profound consequences for the future of healthcare, civil liberties and personal identity. While these developments have been gestating for two decades, they matured and moved centre stage during the pandemic with the massive increase in gene sequencing for tracking Coronavirus. This chapter will examine the social implications of the growing use of genomics and gene-based screening technologies, and analyse how they are contributing to an important shift in how we understand human health. This emphasizes the biological and individualized nature of disease in contrast to the social determinants of health and illness, supporting an increasingly biomedicalized understanding of the human.

A vision of genomic medicine to improve personal and population health

The massive public and private investment in genomics over recent years is inspired by a vision of so called 'personalized medicine', where treatment is guided by a knowledge of an individual's genetic makeup. This is used to assess the risk of disease, design better treatment regimens and predict response to therapy. It is being enabled by the growth of targeted therapeutics and gene-based diagnostics that stratify patients and diseases into discrete sub-populations. This utopian imaginary also aims to improve public health by enabling new forms of population surveillance based on genetic risk profiling and reproductive 'choice' through improved antenatal screening. For its advocates this offers hope for a new kind of medicine which identifies the underlying cause of many diseases as being within the body, rather than in the external environment.

The development of genomic medicine is an international phenomenon, being actively pursued in North America, Europe and East Asia. It is driven by major government investment, as well as novel forms of public-private collaboration, and is best illustrated in the UK, which is establishing itself as a global leader.

Over recent years, a series of large-scale projects have started to develop the core infrastructure for the widespread adoption of genomic medicine in the UK National Health Service (NHS). This has been accompanied by a critique of the NHS as being unresponsive and needing major service transformation through the rapid adoption of new technologies and an innovation culture. At the same time, the NHS has also been recast as a site for the production of new forms of economic value based on the exploitation of digital assets, such as patient data. A pattern of radical service transformation is now visible, with the adoption of genomics and related technologies driven by new organizations and programmes within the public sector that are run on business principles. This growth of genomics is being accompanied by increasing commercialization and digitization of the public sphere and was greatly boosted using gene sequencing to track COVID-19.

One of the most valuable ways to understand the dynamics of a viral pandemic is to track how new variants arise and spread in a population. This can be done in detail by sequencing the viruses that infect large numbers of people. The UK was able to use the genomics infrastructure it was establishing to rapidly start such viral sequencing on a mass scale. This was done by the COVID-19 Genomics UK Consortium (COG-UK), a network of mainly public sector agencies and laboratories that aimed to establish a comprehensive national pathogen surveillance system. By August 2021 it had sequenced over 750,000 viral genomes, the second highest number in the world. In addition to establishing routine population surveillance using genomics, the pandemic also provided an important political opportunity to increase the speed and scale of investment in genomics more generally. This was made possible by greater public willingness to accept population screening, the normalization of gene sequencing, and support for collaboration with a biopharmaceutical industry that had played a key role in developing vaccines.

Genomics England and the rollout of whole genome sequencing

This increased activity was focused around three key programmes: Genomics England was established in 2013 as a company operating as a business within the Department of Health, and delivered the first major whole genome sequencing (WGS) project in the world. This involved decoding every base pair in the genomes of 100,000 UK patients, and was focused on cancer and rare diseases. It provided detailed information on the genetic makeup of patients that can be linked to medical records to help diagnose disease and direct treatment. However, the extent of the clinical utility of this has yet to be fully established. This project laid the foundation for the creation of the NHS Genomic Medical Service (GMS) in April 2020, which seeks to fully sequence between 500,000 and 5 million UK patients by 2024 (Genomics England, 2018), including all children with cancer or those seriously ill with a likely genetic disorder. In addition, there is a commitment to WGS all 500,000 participants in the already established UK Biobank – one of the biggest gene-environment studies ever undertaken. The GMS seeks to bring genomics into the mainstream of UK healthcare by extending access to molecular diagnostics and offering genomic testing routinely to all people with cancer, as well as other 'high risk' conditions such as Familial Hypocholesterolaemia. Most significantly, it will oversee the linking of genomic to other forms of personal medical data.

Integrating genomic data and medical records

A central pillar of personalized medicine is data integration to help stratify patient populations into disease sub-types. The UK has established a massive data infrastructure within the NHS to support clinical decision making and efficient and effective healthcare delivery. The NHS Spine and the Data Processing Services are the main digital platforms and are linked to a series of large-scale databases, that include information on hospital patients and their treatment. The privacy of patient data is supposed to be protected by this being de-identified when processed or shared. In 2021 the UK government announced that it would create a new centralized database (GPDPR) containing the GP records of 61 million patients, that could then be linked to other datasets and shared with researchers and companies on an anonymized basis. This proposal is highly controversial and has yet to be ratified as it operates using presumed consent with patients being able to opt out (Marsh, 2021).

The expansion of genetic screening and testing services

One of the most important potential applications of genomics is in the detection and diagnosis of diseases, as clearly demonstrated by its use in tracking and diagnosing people infected by COVID-19. In addition to the use of genomics to screen newborns, there are proposals to adopt this more widely. The Accelerating Detection of Disease challenge was launched in 2020 as the largest project of its kind in the world and aims to recruit 5 million UK citizens who will participate in a study to improve the early diagnosis of common diseases, such as dementia, cancer, diabetes, and heart disease. It will be organized through a charity – Our Future Health – which will collaborate closely with industrial partners to give public and private sector researchers access to data. Patients will be recruited via routine care and genomic information will be collected and linked to their medical records. These data will be used to create polygenic risk scores, which aim to identify those patients most likely to get certain diseases. These scores will be used to direct therapy and help create new commercial diagnostic products.

The same logic is being applied to COVID-19 , with the international COVID-19 Host Genetics Initiative seeking to identify genetic risk factors that greatly increase an individual's chance of becoming seriously ill. Already companies are developing tests based on these risk scores that can be used to identify people who are most vulnerable. If successful, such technologies could divide the population so that some people would shelter and everyone else would lead a normal life (Callaway, 2021). This could have huge social implications. However, the validity of genomic risk profiling is unproven and remains controversial.

Many aspects of these programmes are still 'in the making' but provide a sense of both the direction of travel and the scale of investment in building these new products and services. However, while they may provide a valuable resource for research and important insights into health and disease, they raise major concerns. Firstly, many of the claimed benefits of genomic medicine have yet to be clearly demonstrated and are largely promissory, with any health gains only being fully realized in the longer term. In particular, the central paradigm of 'screen and prevent', which justifies the surveillance of large populations to detect those at high risk and offer prophylactic therapy, is far from proven. Experience from established screening programmes suggests that this is difficult to achieve. Furthermore, the priority given to this form of high-tech medicine is in danger of distorting healthcare priorities and diverting scarce resources away from less glamorous but more effective interventions that offer immediate benefits at a time of very limited NHS funding. Secondly, there are growing concerns about the increasing commercialization of the NHS. The initiatives described involve close collaboration with industry to undertake sequencing and data analysis. It is not clear that public and private interests are always aligned, and there may be significant conflicts of interest when UK patients and their personal data are being used to test potentially expensive new products. Thirdly, these programmes raise important issues about data access, privacy and security, given the long history of data breaches and unauthorized use of NHS patient records by third parties, including insurance companies (MedConfidential, 2021).

Perhaps the most important consequence of building this genomic and data infrastructure is the associated shift in how we understand human health, and the priority given to a certain type of health future. The focus of genomic medicine is on the internal workings of the body, reinforcing a discourse that the health problems we suffer are biologically determined, with disease and illness being the result of our genetic makeup. This shifts the emphasis onto personal responsibility for leading a healthy life and obscures the way in which many illnesses are the result of social and environmental conditions. In doing so, it naturalizes inequalities in health which are the result of the unequal distribution of wealth. It also justifies much greater surveillance of populations and their stratification into different high-risk groups that are then prescribed prophylactic drug therapy, offered 'lifestyle' interventions, or segregated. This also hides the powerful role that socioeconomic factors play in making people ill while creating new markets for pharmaceutical products.

In the UK the pandemic has only reinforced these narratives about the origins of disease and how we should respond to them. Policy increasingly emphasizes how some people are innately more vulnerable than others, and why we should take individual rather than collective responsibility for managing risk. The investment in genomics both reflects and feeds into this framing of humans as biological subjects. An alternative approach urgently needs to be developed that draws on the insights from this powerful new science, but which sees humans as deeply social and many health problems stemming from the way we organize our societies rather than our genetic inheritance.

References

Callaway, E. (2021) 'The quest to find genes that drive severe COVID', *Nature*, 8 July, https://www.nature.com/articles/d41586-021-01827-w

Genomics England (2018) 'Secretary of State for Health and Social Care announces ambition to sequence five million genomes within five years: news release', 2 October, https://www.genomicsengland.co.uk/matt-hancock-announces-5-million-genomes-within-five-years/

Marsh, S. (2021) 'GPs warn over plans to share patient data with third parties in England', *The Guardian*, 30 May, https://www.theguardian.com/society/2021/may/30/gps-warn-plans-share-patient-data-third-parties-england

MedConfidential (2021) 'Major health data breaches and scandals', https://medconfidential.org/for-patients/major-health-data-breaches-and-scandals/

ELEVEN

Frailty and the Value of a Human in COVID-19 Times

Dawn Goodwin, Cliff Shelton and Kate Weiner

At the outset of the COVID-19 pandemic in the UK, the National Institute for Health and Care Excellence (NICE) produced a rapid guideline (NICE, 2020) for adult critical care that made frailty pivotal to assessments. It aimed 'to maximise the safety of patients who need critical care during the COVID-19 pandemic [and] ... make the best use of NHS resources'. Consequently, 'frailty' became key to both avoiding interventions with the potential to cause harm and, implicitly, to rationing access to care.

What is frailty, how did it become a central construct for making care assessments during the pandemic, and what are the implications of the increasing currency of this concept within healthcare? In this chapter we consider these questions and the multiple constructions of the frail ageing human that emerge from this. We draw on research, media, official reports, and excerpts from an ethnography undertaken on hospital wards before the pandemic.

While frailty has long existed in the vernacular, it has recently emerged as a clinical object around which healthcare is increasingly organized. Clinical conceptualizations of frailty consolidate around a phenotype and a deficit accumulation model. The phenotype depicts frailty as a syndrome comprising characteristics such as weight loss and slow walking speed. The deficit accumulation model has had various incarnations, most recently as the Clinical Frailty Scale (CFS), which positions people on a functional continuum, with assessments aided by descriptors and illustrations.

Efforts to define frailty have transformed it from something 'we know when we see it' into something that can be measured, evaluated, and incorporated into policy. As a clinical concept, frailty is used increasingly for planning and making prognoses, far beyond its origins in geriatric medicine. For example, in England in 2017/18, frailty screening for those aged 65 and over was contractually introduced in general practice, resulting in an automatic categorization of 'frail', 'pre-frail' or 'robust', according to electronic health data (Tomkow, 2020).

Social scientists have repeatedly problematized the language of frailty, showing the term to have multiple and evolving meanings, but nevertheless consistently understood by those who are labelled as frail to be pejorative, stigmatizing and disempowering (Grenier, 2007; Shaw et al, 2018). Concerns are now being raised about the unintended consequences of structuring services around a concept which carries such negative connotations among those who they are designed to serve.

Some suggest that the understanding and operationalization of frailty in general healthcare settings is problematic. Age UK and the British Geriatric Society identified that while specialists in elderly care conceive frailty as a spectrum upon which people can move up and down, other clinicians and older people themselves understand frailty as an irreversible loss of independence (BritainThinks, 2015). Many clinicians also use frailty as a shorthand term for patients with significant needs (Shaw et al, 2018). In short, it is suggested that those who are not specialists in elderly care conceive frailty in a way that incorporates negative lay understandings. Given the enthusiasm for the broader use of frailty assessments in healthcare, this raises questions about the ideas underpinning these assessments, and how they inform care.

Here we would like to contrast ethnographic observations of how frailty can be invoked in clinical decisions, from a (pre-pandemic) study of anaesthesia for hip fracture surgery (Shelton, 2019), with reports of how it has been operationalized during the pandemic. In the following extract, two clinicians consider a challenging case. The patient, Quintin (a pseudonym), is 76 years old and has multiple chronic illnesses. He is on a cardiology ward because of a suspected heart attack, thought to have caused a fall, leading to his broken hip. During his hospital stay he has become confused and breathless. If he doesn't have an operation to fix the hip, he is unlikely to survive because of immobility, pain, and inflammation. However, an operation also carries significant risk.

Briar, the consultant anaesthetist, steps back from the bed and talks quietly – 'He's a bit wheezy … he doesn't pass the end of the bed test, if you know what I mean?' We head to the desk and Briar summarises his train of thought: 'It's a difficult one. … My gut feeling is that he might not survive an anaesthetic. … But if we leave him, he's just going to get pneumonia and die anyway.'

Briar decides to consult his colleague Amos, a critical care consultant. They go to see Quintin together; Amos makes some brief examinations, and they return to the desk where a cardiologist is sat, reviewing Quintin's notes. He hands them to Amos who flicks through and points to an arterial blood gas result. 'He looks like death warmed-up.' Briar looks at the test result – 'he's got type-two respiratory failure'.

'He could have a spinal [anaesthetic]? That doesn't affect respiration' suggests the cardiologist. 'More than you think', Briar retorts – 'the diaphragm keeps working, but the intercostals …' Amos joins in – 'and lying them flat. … They look like death warmed-up, then you shake them up a bit, [cutting] bone, transfusing [blood], the rest …'

Briar draws the discussion to a conclusion – 'We could palliate him? What do you think Amos?' 'I think he'll die in theatre', Amos replies, 'he's frail, using a frame and chair at home, diabetic, heart failure, renal failure, morbidly obese with [obstructive sleep apnoea]. I can't see you making any of that any better.' 'I thought you'd say that', replies Briar. He didn't want to take Quintin to theatre if it would commit him to an intensive care stay that he wouldn't survive.

Quintin didn't have an operation. He died two days following this encounter.

Here two expert clinicians have a dilemma – do they proceed with the operation and risk death in theatre or shortly afterwards? Or decline the operation knowing that Quintin is unlikely to recover without it? They make use of 'frailty', but what do they mean by this? No formal tools are used; instead, they emphasize 'the end of the bed test', a term that suggests clinical assessment is based on how a patient looks 'from the end of the bed'. Yet this term belies the tacit and nuanced assessment based on the clinicians' extensive knowledge and experience.

Briar and Amos are not concerned with frailty in a general sense; most patients with hip fractures are classed as frail. Instead, they focus on Quintin's capacity to withstand stressors that they know will be applied, such as the circulation and breathing changes induced by anaesthetic drugs and the physiological stresses of surgery and critical care. The clinicians therefore explain their concerns with reference to his breathlessness and pallor – implying respiratory and circulatory vulnerability.

Is an insufficiently 'specialist' conception of frailty the problem, with abbreviated assessments assuming an irreversible trajectory and lacking nuance? We suggest that frailty is likely to be understood differentially in different healthcare settings, driven by specific requirements and sufficient for the purposes at hand. In geriatric medicine, frailty assessments are part of a comprehensive assessment and integrated into care planning. In general practice, they are computer-generated and guide preventative interventions, and in anaesthesia and critical care, frailty assessments are specific to the demands of surgical and anaesthetic interventions.

What does this mean for how frailty has been operationalized in the pandemic? Though the development of a frailty-based 'triage tool' was abandoned in March 2020 as concerns about NHS capacity subsided (NHS, 2020), accusations persist that frailty acted as a blunt tool for rationing. Based on investigative journalism, Arbuthnott et al (2020) claim that frail people were excluded from hospital and critical care early in the pandemic. They claim that GPs were required to identify those who would not be admitted to hospital if they developed COVID-19, ambulance and admissions teams were instructed to exclude older people, and care homes were unable to send residents to hospital. These accusations were strongly rebutted by the NHS, which stated that staff were instructed that 'no patient who could benefit from treatment should be denied it' (NHS, 2020). This aligns with the sentiment of the NICE rapid guidance (NICE, 2020) which positions frailty as part of 'a holistic assessment'. However, it has been substantiated that, at the beginning of the pandemic, 'do not attempt cardio-pulmonary resuscitation' decisions were made without patient or family involvement and applied in a blanket fashion to care home residents and frail older patients on GP lists (Care Quality Commission, 2021).

This seems a long way from the nuanced balancing of risk for an individual patient presented in our ethnographic data. Yet, in the absence of detailed ethnographic work, we do not know how frailty assessments played into care decisions made for older patients with COVID-19, the degree to which these involved informal, tacit, and localized assessments by experienced clinicians, or codified frailty scores, or how decisive such assessments proved to be in practice.

In our view, the problematic aspects of frailty do not only arise from lay conceptions and the incomplete understanding of some clinicians. Problems arise when the clinical conception of frailty is used at a policy level. Grenier (2007) contends that in a culture of fiscal restraint, the medical construction of frailty 'is used to ration treatment to those at greatest risk, which places older people in competition for scarce resources'. Yet, she argues, this 'obscures the way that frailty is structured by cumulative disadvantage'. These concerns appear prescient given that in the COVID-19 pandemic, frailty was used to assess older people's risk and, arguably, to ration access to treatment (NICE, 2020; Arbuthnott et al, 2020).

We have shown the multiple potential constructions of the frail older human. Healthcare professionals understand frailty in a way which is relevant to their speciality, the clinical context, and patients' needs. As in Quintin's case, it may be appropriate to acknowledge frailty when considering individual treatment. Wider cultural understandings speak to entrenched forms of ageism, but we should be wary of the implication that only these 'lay' understandings are problematic. Clinical understandings are multiple and can be equally problematic if they shift focus away from structural inequalities. When used at a policy level, frailty may direct attention to the costs of keeping older humans alive and has the potential for large-scale discrimination. Questions remain about the use of frailty assessments during the pandemic, whether the concept of frailty has served or hindered patients in receiving appropriate treatment and care and, ultimately, the rights and respect due to all humans.

Acknowledgements

Cliff Shelton was funded by the National Institute for Health Research (NIHR) for his ethnographic research project

(DRF–2015–08–208). The views expressed in this publication are those of the authors and not necessarily those of the NIHR, NHS or the UK Department of Health and Social Care.

The writing of this chapter was supported by funding from the Foundation for the Sociology of Health and Illness.

References

Arbuthnott, G., Calvert, J., Das, S., Gregory, A. and Greenwood, G. (2020) 'Revealed: how elderly paid price of protecting NHS from COVID-19', *The Sunday Times,* 25 October, https://www.thetimes.co.uk/article/revealed-how-elderly-paid-price-of-protecting-nhs-from-covid-19-7n62kkbtb

BritainThinks (2015) 'Frailty: language and perceptions, a report prepared by BritainThinks on behalf of Age UK and the British Geriatrics Society', 15 June, https://www.ageuk.org.uk/globalassets/age-uk/documents/reports-and-publications/reports-and-briefings/health--wellbeing/rb_june15_frailty_language_and_perceptions.pdf

Care Quality Commission (2021) 'Protect, respect, connect – decisions about living and dying well during COVID-19: CQC's review of 'do not attempt cardiopulmonary resuscitation' decisions during the COVID-19 pandemic', https://www.cqc.org.uk/publications/themed-work/protect-respect-connect-decisions-about-living-dying-well-during-covid-19

Grenier, A. (2007) 'Constructions of frailty in the English language, care practice and the lived experience', *Ageing and Society*, 27: 425–45.

NHS (2020) 'NHS and other professional bodies' response', *The Sunday Times,* 25 October, https://www.england.nhs.uk/2020/10/nhs-and-other-professional-bodies-response-to-sunday-times

NICE (2020) 'COVID-19 rapid guideline: critical care in adults', Guideline (NG159), *NICE,* 20 March, https://www.nice.org.uk/guidance/ng159

Shaw, R.L., Gwyther, H., Holland, C., Bujnowska-Fedak, M., Kurpas, D., Cano, A., Marcucci, M., Riva, S. and D'Avanzo, B. (2018) 'Understanding frailty: meanings and beliefs about screening and prevention across key stakeholder groups in Europe', *Ageing and Society*, 38(6): 1223–52.

Shelton, C.L. (2019). *In search of the 'good anaesthetic' for hip fracture repair: difference, uncertainty and ideology in an age of evidence-based medicine*, PhD thesis, Lancaster: Lancaster University.

Tomkow, L. (2020) 'The emergence and utilisation of frailty in the United Kingdom: a contemporary biopolitical practice', *Ageing and Society*, 40(4): 695–712.

TWELVE

"I've Got People's Spit All over Me!": Reflections on the Future of Life-Saving Stem Cell Donor Recruitment

Ros Williams

Introduction

Like other kinds of tissue donation, stem cell donation systems rely on framing humans as reciprocal, prosocial agents engaging in the communitarian, altruistic act of donation. COVID-19 highlights how, particularly in racially minoritized communities, interpersonal contact in trusted encounters and spaces is a key part of the normative work of encouraging stem cell donation. Thinking through this as an example of how we as humans often need to be co-present to generate affective ties, build new communities and counter forms of inequity and mistrust, this chapter offers an illustration of how the pandemic disrupted this key aspect of being human.

Stem cell donor registries – used to source donor stem cells for treating blood cancers – are currently insufficiently ethnically diverse, meaning that racially minoritized patients in the UK have poorer odds of locating a suitable donor. New donors are registered at in-person events organized to target potential racially minoritized donors. Registrants provide saliva to be tissue typed and placed on the registry. This saliva, full of biotechnological potential, is now also a recognized vector for COVID-19 transmission. As the world reopens, it is hard to imagine that these busy events where saliva-saturated swabs swap hands will restart with the same enthusiasm.

This chapter explores the value of these in-person events in generating registrations amongst racially minoritized people who, importantly, are recruited with *less* success through the digital routes that recruitment work is reliant upon during pandemic restrictions. First briefly contextualizing stem cell transplantation and its attendant racial disparities in access in the UK, the chapter then describes the saliva-orientated work of recruitment. It then reflects on the potential consequences of its cessation.

Stem cell transplantation and racial inequity

Since the 1970s, stem cell transplants have emerged as a well-established and increasingly common treatment (see Passweg et al, 2018). If a patient cannot find a matching donor in their family, their clinician can call upon a global network of registries connecting databases of volunteers to locate donors for their patients. These donors will have initially provided their tissue type via saliva, possibly at a recruitment drive, from which they were logged on a stem cell registry awaiting the rare occasion that they – out of nearly 40 million world-wide donors – might be a patient's genetic match.

Crucially, transplants rely on this genetic match between donor and recipient, and it is understood that unrelated patients and donors are more likely to be a match if they share a similar ethnic background (see Williams, 2018). Given the overrepresentation of white donors on Global North registries, there is therefore a significant health equity issue at play: in the UK, minoritized patients have as little as a 20% chance of finding matches, as opposed to white patients' 69% chance (Smith, 2018).

Such statistics have prompted the emergence of charities seeking to improve registry diversity by running registration drives in places selected based on the chances of having lots of racially minoritized people in attendance. Places of worship, colleges, summer melas and carnivals are just some of the places where such drives might be held (see Williams, 2021). These charities are described as 'crucial' by the registries that provide them with swab kits, and are acknowledged in policy for their central role in improving racially equitable access over recent years (Smith, 2018). They are thus a vital component of the blood stem cell infrastructure, adding thousands of racially minoritized donors to the UK registry each year through in-person recruitment work. When COVID-19 arrived, however, this work stopped.

Event cancelled: the necessary cessation of recruitment work

Altogether, the process of stem cell registration takes three minutes and three saliva swabs. A form collects donor information before the donor rubs three swabs against their inner cheek. These swabs, once saturated with saliva, are collected, sent to a laboratory for tissue typing, and entered onto the globally searchable donor registry. The process, admittedly, can be messy. Take, for example, a Race Against Blood Cancer (RABC) event in the West Midlands, where the charity had been invited to have a stall at a conference in a large exhibition centre. Charity volunteers register 93 new donors, of whom perhaps ten have a minority ethnicity background, a proportion with which the volunteers are very pleased. During a quiet moment, I take this fieldnote of an observation of a charity volunteer:

> '[she] scrunches her face ... like there's a foul smell in the air. She lifts her hands up and, inspecting them, holds them out in front of her like a surgeon avoiding any contact in the moments between scrubbing and gloving: "I'm going to the bathroom to wash my hands. I've definitely got people's spit all over them". She shimmies out from behind the registration desk and dashes off to the bathroom.'

Saliva is a loaded substance (Kragh-Furbo and Tutton, 2017): its attachment with disease transmission has more recently been met with acknowledgement of its biotechnological promise as a non-invasive means of testing for viral particles (for example, a COVID-19 test) or even collecting DNA (for example, ancestry testing). It is how a charity volunteer can be at once disgusted she's got spit on her hands, and thrilled she's managed to collect so many swabs of it. Saliva's loadedness – its capacity to be easily retrieved and then typed in a laboratory – brought us to the exhibition centre, but it was also what would abruptly stop face-to-face donor recruitment work, since respiratory saliva droplets were identified as the primary COVID-19 transmission vector. The same events that stood to save lives through recruiting to registries were also prime opportunities for the spread of disease.

Goodbye, buy-in? The importance of in-person recruitment work

Registries in the UK have for several years invited people to order their own postal swab kits for at-home registration. One needn't go to an event to be swabbed, therefore, but can order a free kit, post their sample, and be registered from home. UK registry Anthony Nolan, however, note that online recruitment is less effective for ethnic minority registration: 'We know that there are limited options for targeting this group online, and have struggled to reach our target for BAME recruitment, so work has begun on partnering offline with organisations, schools and colleges in diverse areas' (Anthony Nolan, 2019: 14), gesturing to the sense that – particularly for some racially minoritized people – there is value in physical co-presence when trying to engage people in donation.

Nonetheless, just as COVID-19 resigned many to homeworking facilitated purely by digital platforms, recruitment work moved online too. The African Caribbean Leukaemia Trust (ACLT) hosted social media interviews with Black celebrities, from Paralympian Ade Adepitan to entertainer Alesha Dixon, where charity organizers interviewed them about their lives, interspersing this with why and how audiences (hopefully predominantly Black) might register from their own homes. Charities so adept at face-to-face work had instead to encourage people to order swabs in their own time and join the registry from home, a process noted to be less effective for recruiting minorities.

Charities like ACLT and RABC are seen in policy as key conduits between registries and racially minoritized communities, able to generate trust because of their distance from statutory actors, and to make the recruitment message more legible to racially minoritized audiences by adopting a particular mode of communication (see Williams, 2021). This work can be challenging, with those undertaking it acknowledging that there is good reason why some racially minoritized people may say 'no' to requests to participate. For example, the following fieldnote captures a fragment of my conversation with a recruitment co-ordinator at RABC. He describes how growing up as a Black male in London, surrounded by Black friends, he encountered a reaction that is also common amongst the racially minoritized people he approaches to recruit:

> ' "People take your stuff. The government takes this, the government takes that. It's distrust. What will they do with my DNA? With my details?" I asked if this reaction didn't make some sense, given the context in which those opinions will have been formed. "100% I get it, There's no part of me that doesn't get it".'

The same coordinator thus regularly goes back to the same places, both recruiting new donors, and developing relationships with local contacts, to gradually persuade those who are less inclined to register because of the 'distrust' they may have, particularly of statutory systems. Vitally, though, this work is done in-person. In an interview with an ACLT leader, I ask him about the effects of all the cancelled events because of COVID:

> 'The amount of stem cell donors that we've recruited … will be minuscule in comparison to what we would normally be doing in terms of hundreds, if not thousands … when you're in physical face-to-face, there is a much stronger buy-in because they're feeding off your body language, your terminology, your presence and you're there afterwards for reassurance if they want to maybe talk to you or the team. So there is a strong buy-in at the time, you've got them hooked … emotionally, physically, so you've got a stronger chance of them saying, "yes, I'm going to do this now, this is my moment", rather than going away and thinking about it. Online, yes, you lose that first-hand connection.'

In-person events, as I've argued elsewhere, rely on physical proximity to develop a personal connection between charity workers and would-be registrants that mobilizes an *ethico-racial imperative*, a sense of donation as a morally good course of action within a racialized community (Williams, 2021). But events are also, as the extracts gesture towards, spaces where highly embodied reassurance work is undertaken. Body language, presence and talk combine as volunteers attempt to foster sufficient will in potential donors, perhaps to overcome a concern about where their data will travel, to register. You "lose that first-hand connection", as the interviewee describes it, with online recruitment work.

Conclusion

Alongside other face-to-face interaction, an important but underacknowledged kind of work – the in-person donor recruitment drive – was also lost to COVID-19. These events serve a crucial purpose, bridging many racially minoritized donors with a system that, in its current state, isn't offering sufficient provision for racially minoritized patients. The evident value of in-person recruitment, particularly for bolstering some donors' trust in the recruitment process, cannot be disregarded. Though we cannot know, future patients, who might have otherwise benefited from stem cell transplants from donors that did *not* join the registry during the pandemic, might now not receive their life-extending transplants. It is something of a cruel calculus that the very actions designed to save lives – lockdowns and restrictions on movement – stymie work that has its own life-saving potential.

More broadly, though, events like these, which rely on being physically together, highlight how we as humans often rely on co-presence to build community and affective connection. Such conditions are, as we see in the example of donor recruitment, also central to redressing inequity. We might exercise some caution, then, at the suggestion that things may simply 'move online', for a key part of being human risks being lost.

Acknowledgements

This project was funded by a Wellcome Trust research fellowship Grant No. (212804/Z/18/Z).

References

Anthony Nolan (2019) Annual report, https://www.anthonynolan.org/sites/default/files/2021-02/1929cm_annualreport2019_digital_aw4.pdf

Kragh-Furbo, M. and Tutton, R. (2017) 'Spitting images: remaking saliva as a promissory substance', *New Genetics and Society*, 36(2): 159–85.

Passweg, J.R. et al (2018) 'Is the use of unrelated donor transplantation levelling off in Europe? The 2016 European Society for Blood and Marrow Transplant activity survey report', *Bone Marrow Transplantation*, 53(9): 1139–48.

Smith, E. (2018)' Ending the silent crisis: a review into Black, Asian, mixed race and minority ethnic blood, stem cell and organ donation', *Sheffield Street Company,* 18 May, https://www.nbta-uk.org.uk/wp-content/uploads/2019/04/BAME-Donation-review-29.5.18.pdf

Williams, R. (2018) 'Enactments of race in the UK's blood stem cell inventory', *Science as Culture*, 27(1): 24–43.

Williams, R. (2021) ' "It's harder for the likes of us": Minority ethnicity stem cell donation as ethico-racial imperative', *Biosocieties,* 16(4): 470–9.

THIRTEEN

Science Told Me (But I Couldn't See Its Point)

Rod Michalko

In the following, I invoke the experience of disability, my blindness, and of ageing, and I will show how scientific responses to COVID-19 have rendered the combination of the two not only as a site of vulnerability, but also as a life barely worth living. I will explore how vision together with its sister, the virtual, have diminished touch. Touch has become a method for vision to elevate itself to the 'master sense'. Vision is thus regarded as the best way to experience the inter-human, constituting blindness, with its 'master sense' of touch, as not merely second best, but as beneath notice by and irrelevant to the human project. Finally, I will show how blindness and old age co-mingle and how science has used COVID-19 as the occasion to emphasize the diminished humanity of both.

Austin Clarke[1] once told me that he was 'getting old' and that he 'hated it'. And now, I too am getting old and like Austin, I hate it. When you think about it, *really* think about it, there isn't anything good about getting old. And, blindness doesn't help old age get any better.

One day at a time; live life one day at a time. This is an admonition that old age often casts our way. Harsh as it is, it does seem as though it is our only choice; not that anyone, regardless of age, can live life more than one day at a time. But, in old age – this becomes a conscious choice. Living today as though it were the last day of your life takes on an eerie poignancy in old age. This is captured very well in an old joke; a person of 85 years is asked, "How do you start your day?" They reply, "I read the obituaries in the newspaper. If my name does not appear there, I get out of bed."

Perhaps the situation is not as drastic as this, but only perhaps. The sense of time being short and quickly passing, as well as the dread of wasting what time remains and the need to grab it as it passes by, haunt us in old age. A lament – I won't get those two hours back – takes on the brilliance of a clear bitterly cold winter's day and has the same humbling effect. Time, we realise in old age, is indeed precious.

Whatever else old age is and however it is experienced, it is our final lap; it is the last time we will have with and around this thing, this place, this space that we call – my life. For some in old age, taking this lap as quickly as possible seems the only choice; for others, slowly, moving around this final lap seems necessary. Whatever the choice, our final lap is taken in our present and with our past, not behind us following, but at our side taking every move of our final lap with us. Our future? Who knows? One thing is certain, though; we are so close to our future in old age, so intimate with it that we know it is there, right there, and it will become our present. ... At any time.

Since I am writing this in the midst of COVID-19, it seems appropriate to reflect on the malady of old age in the midst of what is going viral. Despite the fact that COVID-19 has gone viral, it must be stated, obvious as it is, that the malady of old age has beaten COVID-19 to the punch. Old age has gone viral long ago, long before COVID-19. COVID-19 has not nor will it ever make as many people ill as has old age, nor will it kill as many. What COVID-19 has done to old age is to allow people to transform those of us growing old into 'the elderly' and then, into the 'vulnerable'. This transformation is, of course, not something new; it has occurred long before COVID-19, too. What COVID-19 has done is to give scientific and medical permission to our families and friends, to our communities, to our countries, even to the world – permission to legitimise their treatment of us as vulnerable and to feel good about doing so. And this, as so many of 'the elderly' die in those very facilities euphemistically called 'long-term care'. Along with the legitimisation of the treatment of 'the elderly', COVID-19 has also exposed, as though for the first time, the severely inadequate and inhumane treatment 'the elderly', our most vulnerable, receive in those long-term facilities labelled with the misnomer … care.

Perhaps most devastating, is that COVID-19 has given permission to all of society to 'see' this inhumane treatment as something COVID-19 has brought along as a travelling companion in its viral journey. It is better for all society to understand the deaths of so many people in long-term care facilities as a result of COVID-19, and not as one more instance of how our society thinks of and feels about growing old. Most troubling is that the permissions doled out by COVID-19 to treat 'the elderly' as 'vulnerable' comes uncomfortably close to the permission to cull … the herd.

It's hard to believe that it was more than a year ago in late January 2020 that the first case of COVID-19 was detected here in Toronto. This was followed by a series of 'lockdowns' with the exception of facilities deemed essential; hospitals and medical emergency services, police and fire services, pharmacies, grocery stores and, of course, beer and liquor stores were allowed to keep their doors open. People were asked to stay at home and to leave only in order to use the essential services. Social distancing, now merely a habit, became the mantra of the day – stay six feet away from each other.

COVID-19 became the tyrannical leader of nearly the entire globe and ruled with the motto 'obey or get sick and die'.

Two other rulers emerged shortly after COVID-19 took over – science and vision.

Science had for a long time been seen as the source of all 'real' knowledge; but COVID-19 elevated its status. Science was now revered and world leaders, with a few exceptions, followed science, bowing in its wake. "We're following the science and locking down", became the cry of these leaders. Science appropriated the role of Decision Maker for all of us. Scientists dropped the personal pronoun 'I' from their vocabulary. They referred to themselves as we, a euphemism for science. The 'we' took the word 'evolve' to another level as well. It became a code for, 'we don't know what we're doing', except scientists expressed it as, 'the science around COVID-19 is evolving'.

Like science, visuality and its companion, the virtual, were also elevated to lofty heights by COVID-19 and, like science, they too already held a lofty position in the contemporary world. The sense of touch was shoved down to the low level visuality thought was its rightful place. Any ground that touch may have gained in our times was quickly swept from under its feet. Science and visuality forced touch to inhabit the animal and dangerous aspect of human life, something it had always represented. Touch held a quasi-taboo status in our culture and this status is now full-blown.

We no longer touch one another; we keep our distance. The only contact we now have with people, other than our immediate bubble of family and friends, is visual contact. Eye contact has expanded as exponentially as has COVID-19 itself. We still hug; but, not really; we do so virtually as we do much of everything else. The virtual, the 'non-real' aspect of visuality, is now the first lieutenant of sight; it now occupies the space to the right of the re-anointed 'master' of the sensorium – Sight. Touch kneels humbly at their feet.

Along with touch, blindness too has been relegated to the heap of disposable perceptions by science. We know, proclaims public health professionals, that one way COVID-19 is transmitted is through touch. If you touch people or surfaces, do not touch your face. We know that this will transmit the virus from others to you. Oh – science has evolved – you can now touch surfaces. But, don't breathe on others and don't let others breathe on you – if you see them coming. And still, we don't touch. 'We' has made it 'virtually' impossible for us to touch others, things, and even ourselves. Experiencing the world and one another visually, virtually, has become the preferred option of experience, the one that 'we' dictates. The way blind people experience the world and others, and the legitimacy of this experience, has plummeted to new levels of non-acceptance and often of despair as well.

Add to this situation the malady of old age and we have another crisis, one as genuine as that of COVID-19. 'The elderly', we are reminded time and time again by 'we', are particularly vulnerable to COVID-19, especially those with underlying conditions. 'We' sometimes calls this comorbidity, the presence of two or more conditions or diseases that may cause death; heart disease, diabetes, are two such conditions. What is striking, though, is that 'we' suggests that either one of these conditions when paired with the malady of old age, is comorbidity. Old age is a cause of death; the cause of death inscribed upon the death certificate, however, will read 'natural' causes. It is 'natural' to die of the malady of old age; it is a death attributed to natural causes.

No wonder 'we' is so concerned with the vulnerability of 'the elderly' in these times of COVID-19. If we get it, we die. Best stay at home. Long-term care facilities, on the other hand, appear to be comorbid with the malady of old age. Along with diabetes and heart disease, long-term care appears to be one of those underlying conditions suffered by some with the malady of old age. Not surprisingly, the science, particularly epidemiology, is evolving and 'we' know things about COVID-19 'we' didn't know a few months ago. What 'we' know now with regard to COVID-19 and old age is that 'the elderly' are now less likely to get the virus than are those under 50 years old. Since bars, restaurants, parks began to reopen with the opening of the economy, the under 50 age group are the ones going to these places and socialising, sometimes without social distancing and often without masks. This, the epidemiologists tell us, is what is causing this age group to be more likely to contract COVID-19 than are 'the elderly'.

Still, the news isn't all good for 'the elderly'. If they get it, 'we' tells us, they are more likely to die than the under 50s.

After all this, nothing has changed, really; I'm still growing old. And I still hate it. COVID-19 has not changed this. What has changed is wondering how long I have to wait for 'we' to distribute the vaccine for COVID-19. It's almost as though the malady of old age and COVID-19 are in some sort of macabre race; each trying in an equally macabre fashion to be the first to push old age out from the infinity of the future into the finality of its finitude.

So, like all those with the malady of old age, I wait. I wait for 'we' to move from evolving to evolution – to a vaccinated world population. I wait for this in the rubble left in the wake of COVID-19. I wait; wondering which is better – going out with COVID-19 or going out in COVID-19? – or, perhaps I'm wondering which is worse.

Note

[1]   The following quotations are taken from the late Austin Clarke's award-winning short story, They Never Told Me (Clarke (2013) *They Never Told Me: And Other Stories*, Toronto: Exile Editions).

PART IV

Human Futures

What will it mean to be human in the future? A future in which different kinds of humans can flourish together is possible, but so too is one of social and ecological breakdown. Will we live alongside robot companions and genetically edit the next generation? Can we rethink how the economy works or how we educate our children? This section considers not what will be, but how we imagine what could be, examining the future as something co-created between science, technology, politics and society in the here and now.

FOURTEEN

Where Will an Emerging Post-COVID-19 Future Position the Human?

Keren Naa Abeka Arthur, Effie Amanatidou, Stevienna de Saille, Timothy Birabi and Poonam Pandey

Introduction

Despite long-standing warnings about the likely emergence in the near future of a novel coronavirus to which humans had no immunity, the pandemic caused by COVID-19 caught most countries unprepared. As of 18 April 2021, the world had recorded three million deaths due to COVID-19.[1] The series of lockdowns adopted by most countries in an attempt to slow down the spread of the virus resulted in about 255 million full-time job losses in 2020 (ILO, 2021) many of which are likely to become permanent even after the pandemic is brought under control. Controlling COVID-19 has further challenged the fundamental human right of freedom of movement and ushered in a range of new methods of surveillance and control of public behaviour. It also underlined long-standing inequalities, with the burden of illness falling disproportionately on already disadvantaged groups, who tend to be more precariously employed (Warjri and Shah, 2020).

As worrisome as these trends are, it is also possible to imagine a future where COVID-19 leads us towards something more equitable, more humane and beneficially globalized (as well as better preparation for pandemics to follow). Although the future is deeply uncertain, and as of the time of writing the pandemic is far from over, we have also identified some positive trends which are worthy of discussion. We argue that the pandemic has shown us the possibilities which might be derived from shifting our focus from an economy based on always increasing monetary exchange of goods and services, as measured by gross domestic product (GDP), to an economy focused on increasing social resilience and wellbeing. Key to this is a shift away from our present obsession with market-based indicators and a movement towards a growth–agnostic approach to innovation. In previous work, we have called this 'responsible stagnation' (de Saille et al, 2020), a term which could also describe the global halt to economic activity which occurred in March 2020 as the only means of stopping the spread of COVID-19. In this chapter we highlight some trends towards growth–agnostic, social innovations that occurred during 2020, focusing on human flourishing rather than economic growth.

Patterns before COVID-19

Since World War II, increasing emphasis has been placed on outcomes as an indicator of economic health. This led to a fast-moving human lifestyle, and a focus on increasing throughput while driving down production costs (that is, wages). By determining what to produce and when to sell, the forces of supply and demand were framed as solely capable of effectively allocating resources (Degutis and Novickytė, 2014), and economic aspiration was framed as the centre of human activity. This was portrayed through the abstraction of Economic Man – the self-interested breadwinner, acting perfectly rationally based on perfect information, serving and served by perfect markets.

However, human migration is also dictated by the market, with more and more people choosing to locate themselves in cities, not only for work but also access to schools, hospitals, and retail/leisure activities, as rural areas decline. Affordable housing in cities, however, has become increasingly difficult to find, and service sector employees now make so little, even on full-time wages, that governments must step in with subsidies in order for them to survive. Thus, markets have shown themselves to be the opposite of efficient (Farmer et al, 2012) and inequality, exclusion and depletive resource use have been the result of the demand for perpetual growth.

The economy versus the human during COVID-19

The emergence of the coronavirus pandemic led to a global grappling with the tension between preserving human life and preserving economic activity. Countries that leaned toward the economy in 2020 (USA, UK, Sweden, India, Brazil) have unsurprisingly done less well healthwise in comparison to those who took a more severe and consistent lockdown approach (New Zealand, Australia, South Korea, Vietnam), however, in no country has the national economy fared well.

Nevertheless, despite the economic stagnation of 2020, certain patterns appeared which illustrate opportunities to remake society in more responsible, sustainable, and equitable ways. Social innovation, always stimulated when the market fails, developed new ways of improving access to valid information, new telehealth services and microfinance loans, and scaled up community-based COVID-19 screening.[2] These makeshift solutions tended to use or require less resources (what is known as jugaad innovation), and to emerge bottom-up and organically through community involvement, rather than through top-down initiatives. For example, the Kaundu Community-Based Health Insurance Initiative reduced out of pocket expenditure in a rural district in Malawi through a localised health insurance scheme, and Chipatala Cha Pa Foni, a national nurse-led health information call centre, provides health information free of charge (van Niekerk, 2020).

Driven by care needs, social innovation also combined with patterns of reverse rural migration. Moralli and Allegrini

(2021) note new attention for Italian inner regions due to the COVID-19 crisis in that country, with the city as a place of dense sociality contrasting with an idyllic vision of rural areas as romantic, healthy, and safe. New forms of care showcasing solidarity over monetary profit emerged as part of this migration. For example, a voluntary service of home delivery for the elderly was activated in the Puglia Region, while a hotel in Brescia province, managed by a cooperative that promotes social and economic integration of asylum seekers, hosted doctors and nurses working in the local hospitals during the peak of the pandemic.

Probably the most ubiquitous jugaad care innovation, early in the pandemic, were the numerous how-to videos circulating on social media illustrating how to make masks out of items that would be on hand during a lockdown, such as old button-down shirts, bandanas and hair elastics. This idea of doing new things with what was at hand furthered existing trends in visible mending, upscaling old furniture, and urban gardening. The COVID-19 crisis also forced some governments to adopt social innovations that had been unthinkable in February 2020, such as basic income schemes, rent and mortgage holidays, and eviction moratoria for both business and private tenants.

However, the most visible trend has been digitalization of the physical spaces that once defined a very broad spectrum of human interactions. Education, health, work and culture have all been given over to digital solutions which have profoundly changed the norms of how we engage each other, forcing new demands on existing local communications infrastructure. In Ghana, for example, new investments were channelled into radio and television in order to solve communication problems in education, as these offer a wider coverage than broadband in that country. During the first lockdown, artists gave Zoom concerts while theatres, opera houses and dance companies streamed their performance archives for free. Education, while homebound, simultaneously became more globalized as people tuned in to seminars and talks given by colleges, libraries, bookstores and charitable organizations from all parts of the world. Such activities represented a new and most welcome kind of global responsibility-taking for the maintenance of human social cohesion and mental health.

(Re)-positioning the human in a post-COVID-19 future

Unfortunately, the market mentality has largely prevented these care-oriented innovations from becoming permanently supported, certainly not by governments desperate to restart growth. While the initial lockdowns served as a sudden roadblock for the driving forces of the last 30 years, old systems are quickly being revived. Innovation driven by limited resources, a more common practice in the Global South, helped to position the human (rather than growth) at the centre of pandemic economies, but this can only continue if supported by new norms demanding economic policies which prioritize health and welfare over GDP.

The digitalization of human interaction effectively allowed the world to continue to function, despite the appearance of coming to a halt. The ability to continue work activities online meant that for many, place of residence was suddenly irrelevant to employment, and the debate over rural investment in broadband infrastructure took a vigorous new turn. Wang et al (2020) describe this as a shift from a factory paradigm characterised by rigid and formalised norms, to a hypermobility paradigm comprising a nomadic work life with no restrictions on where and when knowledge workers work.

But the Digital Nomad is not really the new figure of market-driven choice and limited tensions between work, leisure and mobility it claims to be (Cook, 2020). COVID-19 has also created new forms of digital inequality, made clear by the reliance on online learning during the extended periods of lockdown – a problem not limited to, but of crucial concern in, countries where access to basic education is already neither free nor guaranteed (Azubuike et al, 2021). We cannot take for granted that seemingly beneficial COVID-driven reversals have not also created new pressures on people, in particular loss of community due to more isolated working patterns, and uncomfortable coexistence of work, school and family life in constricted spaces. Replacing Economic Man with the Digital Nomad would change little; the latter trend seems even more likely to have negative implications in the form of ubiquitous precarity, while factors such as gender, age, family size, accommodation arrangements, and reliable communications access, will all have a critical impact on whether this represents increased pressure or new opportunities for human flourishing.

At this moment, the post-COVID future is still somewhat open. There is still an opportunity to consider what the last years have taught us, and to build on these realisations by continuing to demand – as in the summer protests of 2020 – that we as humans do better by the planet, and by each other. The pandemic can be considered a double-edged sword, one that clearly showed the inequities inherent in what has essentially been a global experience. But we have also seen that more, much more, is possible than we have been led to believe. Care became a central value at the beginning of the pandemic – we stayed home, we spread out, we looked after each other in completely new ways. While the end of the pandemic cannot yet be predicted, we still have opportunities for increasing equity through a post-COVID-19 convergence of market-driven *and* social-driven innovations, aimed not at 'going back to normal' or ramping up GDP, but at extending the survival tools and relational skills developed during the first lockdown more broadly. We must use the pandemic to build a better 'normal' of our own choosing – before a normal we cannot survive is built for us.

Notes

[1] Johns Hopkins Coronavirus Resource Center, https://coronavirus.jhu.edu/
[2] World Economic Forum, https://www.weforum.org/agenda/2020/03/how-social-innovators-are-responding-to-the-covid19-pandemic/

References

Azubuike, O.B., Adegboye, O. and Quadri, H. (2021) 'Who gets to learn in a pandemic? Exploring the digital divide in remote learning during the COVID-19 pandemic in Nigeria', *International Journal of Educational Research Open*, 2: 100022.

Cook, D. (2020) 'The freedom trap: digital nomads and the use of disciplining practices to manage work/leisure boundaries', *Information Technology and Tourism*, 22(3): 355–90.

de Saille, S., Medvecky, F., van Oudheusden, M., Albertson, K., Amanatidou, E., Birabi, T. and Pansera, M. (2020) *Responsibility Beyond Growth: A Case for Responsible Stagnation,* Bristol: Bristol University Press.

Degutis, A. and Novickytė, L. (2014) 'The efficient market hypothesis: a critical review of literature and methodology', *Ekonomika*, 93(2): 7–23.

Farmer, R.E., Nourry, C. and Venditti, A. (2012) *The Inefficient Markets Hypothesis: Why Financial Markets do not Work Well in the Real World,* Working paper 18647, Cambridge, MA: National Bureau of Economic Research.

ILO (International Labour Organization) (2021) *ILO Monitor: COVID-19 and the World of Work*, 7th edn, Geneva: ILO.

Moralli, M. and Allegrini, G. (2021) 'Crises redefined: towards new spaces for social innovation in inner areas?' *European Societies*, 23(suppl 1): S831–S843.

van Niekerk, L. (2020) 'COVID-19: an opportunity for social innovation?' International Health Policies Blog, 27 August. https://www.internationalhealthpolicies.org/featured-article/covid-19-an-opportunity-for-social-innovation/

Wang, B., Schlagwein, D., Cecez-Kecmanovic, D., and Kahalane, M. (2020) 'Beyond the factory paradigm: digital nomadism and the digital future(s) of knowledge work post–COVID-19', *Journal of the Association for Information Systems*, 21(6): 1379–401.

Warjri, L. and Shah, A. (2020) *India and Africa: Charting a Post-COVID-19 Future*, New Delhi: Observer Research Foundation.

FIFTEEN

(Genome) Editing Future Societies

Michael Morrison

> Marginalized communities (including those who are low-income and those who come from historically disadvantaged communities of colour) are often unable to access the benefits of science and technology, yet may be disproportionately subject to the harms.
>
> Shobita Parthasarathy[1]

The global COVID-19 pandemic is relevant for gene editing primarily because of what the pandemic reveals about the politics of technology. Technologies, in their design, their function, the systems through which they are distributed and made available, owned, controlled, and regulated, are never neutral. They impact different people, and different groups in society, differently. The distribution of benefits and harms from technologies often, though not inevitably, mirrors existing patterns of inequality and vulnerability.

The social determinants of health are well known in public health, if often neglected in health policy. However, coronavirus can be thought of as a 'vulnerability multiplier'. People already living precarious existences, whether because of poverty, discrimination, pre-existing illnesses, or intersectional combinations of these and other factors, are more likely to be exposed to COVID-19, to suffer worse outcomes when they contract the disease, and to have a harder time dealing with precautionary measures such as social distancing, even when they do not contract the virus.

Importantly, the inequality-exacerbating effects of the pandemic are not limited to the virus itself, but extend to the various technologies deployed to manage, treat or protect people, from masks and PPE, to Remdesivir and other drugs that manage symptoms of COVID-19 infection, and of course, the COVID-19 vaccines. Historically disadvantaged communities, such as African-Americans in the US, may be further disadvantaged by limited and inadequate access to the vaccination programme and uncertainty about trustworthy sources of vaccine safety information (Dembowsky, 2021). In the UK, it has been reported that concerns raised by BAME NHS staff about inadequate and unsafe PPE were ignored by the management at some hospitals (Harewood, 2021).

Access and fairness issues manifest at the global scale as well. Access to Remdesivir has been beset by high prices and uneven access within and between countries (Boodman and Ross, 2020). The current distribution of doses of the major COVID-19 vaccines has been massively skewed in favour of the world's richer nations, which have received an estimated 86% of all COVID-19 vaccine produced to date (Collins and Holder, 2021). This reflects an innovation ecosystem that values strong intellectual property rights and the economic interests of the pharmaceutical industry over global public health or reducing inequality. Plans for another 'technical fix', in the form of vaccine passports for travel and access to essential services, have similarly been criticized for their potential to exacerbate and amplify existing inequalities and disparities within and between countries.

These effects arise partly because technologies – and policies – are often designed with a particular 'user' in mind. This imagined user typically reflects the majority characteristics of the groups developing technologies or promulgating policies (for example, on vaccine rollout or PPE). In Western countries this usually means imagined users are white, male, able-bodied, heterosexual, and middle class. Other groups may be acknowledged as needing additional or special measures, but the very fact that they are positioned as 'other' than the imagined 'normal' or typical user is still inescapably discriminatory. This insight is not new, nor is it unique to COVID-19. It is simply that the global nature of the pandemic has made it more abundantly clear and harder to ignore than was previously the case. In this chapter, I will apply this lesson to gene editing, and specifically to its most controversial application, gene editing of human embryos to produce inheritable mutations that will be passed onto future generations.

The idea of heritable or 'germline' genetic modification is not new. If genetic changes can be made to an embryo, or to gametes (egg or sperm) which then combine to form an embryo, those changes will be present in every cell that derives from that embryo; that is every cell in the developing, and eventually adult, organism, including its own gametes. As a result, the changes can also be passed onto that organism's offspring and so on, through the generations. Applied to humans, the idea is particularly controversial, because it raises the eugenic prospect of parents, the state, or some other entity, being able to select what kinds of future children they consider suitable to exist.

In the 1980s two key distinctions were proposed to guide the use of human gene editing. Firstly, genetic modification of cells in the adult body ('somatic' modification) was distinguished from inheritable modifications (made to embryos or gametes). Secondly, genetic modification to treat existing (genetic) diseases was distinguished from the genetic modification of humans with the intent of enhancing human traits above normal levels or adding wholly new capabilities (United States President's Commission, 1982). In practice both options are frequently merged, since at least one recurring eugenic vision of germline gene editing views it as a means to create 'better humans' by enhancing future generations (cf Savalescu, 2001). In any case, both germline modification and human enhancement were deemed ethically problematic by the 1982 President's Commission and not to be pursued. This stance was rapidly adopted by bioethicists and attained legal force in many jurisdictions (Isasi et al, 2006). However, what was a technically challenging prospect in 1980 is much more feasible in 2021, with new 'gene editing' tools now widely available.

Gene editing refers to a set of molecular tools and techniques that can be used to modify DNA. As with much of molecular biology, gene editing tools have names such as Zinc Fingers, TALENS, or CRISPR–Cas9 that do not immediately convey much information or meaning to non-specialists. In essence, all gene editing tools operate in a similar way; they contain a programmable 'targeting' domain that can be designed to find and attach itself to a particular sequence of genetic material in a living cell. Once the specified target DNA sequence is found, a second part of the tool, which can be thought of as a 'molecular scissors', can cut out that particular piece of the DNA, replace it, or change its content (for example, changing an 'A' to a 'T' in the genetic code). This ability to design gene editing tools to target particular sequences of genetic material is widely regarded by scientists as giving them greater control and allowing them to be more accurate than was possible with older techniques of genetic modification.

Most of the excitement, and indeed hype, around gene editing has arisen because of one particular tool, known as CRISPR–Cas9. CRISPR stands for Clustered Regularly Interspaced Short Palindromic Repeats, which describes the programmable targeting part of the tool, while Cas9 is an enzyme (the 'molecular scissors') which cuts or otherwise modifies the DNA. First reported in 2012, CRISPR is widely championed as being a considerably quicker, easier, and more efficient means of editing genetic material than any previous method of genetic modification (Ledford, 2015).

In 2018 a Chinese scientist, He Jiankui, reported the birth of two babies from implanted embryos whose DNA had been genetically modified using CRISPR–Cas9 gene editing, moving the heritable genetic modification debate from possibility to reality and putting the discussion inescapably back in the scientific, and ethical, spotlight (Morrison and de Saille, 2019). Jiankui's reported motivation for conducting the embryo editing was to confer resistance from HIV to the children by making genetic changes that make it harder for certain strains of the AIDS virus to infect human cells. Although details are hard to verify, it seems clear that Jiankui's experiment did not achieve the edits he sought to make. How the genetic changes that he did produce will affect the children now, and over the course of their lives, remains uncertain. Nonetheless, a number of authors have more recently floated the idea of using genetic modification to try and confer resistance to coronavirus infections and enhance human immunity to COVID-19 (Adashi and Cohen, 2021; Germani et al, 2021).

While germline modification would be only one way to achieve this, and all authors agree there is still work to be done in understanding the biology of human immune responses to COVID-19, the technical tools now exist to make human (germline) gene editing a realistic option to protect against this or future pandemics.

However, while genetic technology has moved on from the 1980s, in many ways the debates about germline gene editing have not. In most western, liberal democracies the dominant legal and political model is underpinned by the figure of the rational, responsible, decision-making individual citizen. This figure is strikingly similar to both the self-serving *homo economicus* of economic theory and the autonomous individual human subject of moral philosophy and bioethics. A core limitation of all these conceptions of personhood is that they envisage people, whether citizens or subjects, as essentially interchangeable and wholly discrete units. This has led debate to focus on abstract questions of the moral or legal rights and wrongs of germline gene editing applied to these interchangeable subject–citizens, now and in future generations.

COVID-19 is but the latest, starkest, illustration that humans are deeply, inescapably interconnected, and that such connections are crucial to our individual and collective wellbeing. Moreover, equality, whether in law or in the commitment to recognizing the moral value of all human beings, does not mean that we should view people as essentially equivalent or interchangeable. COVID-19 has illustrated how social characteristics – age, gender, class, race, sexuality, (dis)ability, health, wealth, geographical location, and social status – directly affect whether, and on what terms, people have access to technology, and how different people and groups experience the harms and benefits of technology. Consider, for example, how prophylactic gene editing for coronaviruses might work in practice; would entry to certain countries be restricted to humans who could prove they had been edited? What care would still be provided to those who could not, or chose not to, access gene editing?

Beyond the coronavirus example, with its capacity to inscribe societal prejudices and biases into the genetic makeup of future generations, heritable gene editing has the potential to drive a genetic homogenization among those parts of the world that have access to such technology, while further marginalizing the needs and priorities of those who have no hope of accessing such technology. If discriminated-against characteristics become 'edited out' of wealthy populations, social and political support for diversity and difference are likely to be further eroded, once difference becomes framed as an avoidable 'choice'. As a result, germline gene editing must surely rank as one of the technologies with the greatest potential to replicate and exacerbate inequalities in the 21st century; the ultimate vulnerability multiplier.

However, if the lessons of coronavirus are heeded, we must recognize that the ethical debate over germline gene editing must abandon its preoccupation with matters of individual self-determination and choice, and focus instead on the (collective) needs, values and voices of those humans most likely to be exposed to the harms of a global market in germline gene editing. This might begin with the disabled, and in particular groups such as people with autism or hereditary deafness, where at least some of those affected find value in being (genetically) the way they are. The issue at stake must be whether, not when or how, to employ the technology.

Note

[1] The opening quote is taken from the testimony of Professor Shobita Parthasarathy before the US House Appropriations Subcommittee on Energy and Water Development, and Related Agencies on 25 February 2021.

References

Adashi, E.Y. and Cohen, I.G. (2021) 'CRISPR immunity: a case study for justified somatic genetic modification?' *Journal of Medical Ethics,* http://dx.doi.org/10.1136/medethics-2020-106838

Boodman, E. and Ross, C. (2020) 'Doctors lambast federal process for distributing COVID-19 drug Remdesivir', *STAT News,* 6 May, https://www.statnews.com/2020/05/06/doctors-lambaste-federal-process-for-distributing-covid-19-drug-remdesivir/

Collins, K. and Holder, J. (2021) 'See how rich countries got to the front of the vaccine line', *New York Times,* 31 March, https://www.nytimes.com/interactive/2021/03/31/world/global-vaccine-supply-inequity.html

Dembowsky, L. (2021) 'Stop blaming Tuskegee, critics say, it's not an excuse for current medical racism', *NPR,* 23 March, https://www.npr.org/sections/health-shots/2021/03/23/974059870/stop-blaming-tuskegee-critics-say-its-not-an-excuse-for-current-medical-racism

Germani, F., Wäscher, S. and Biller-Andorno, N. (2021) 'A CRISPR response to pandemics? Exploring the ethics of genetically engineering the human immune system', *EMBO Reports,* 22: e52319.

Harewood, D. (Presenter) (2021) 'Why is COVID-19 killing people of colour?' television broadcast, 2 March, London: TwentyTwenty Productions for BBC One.

Isasi, R.M., Nguyen, T.M. and Knoppers, B.M. (2006) *National regulatory frameworks regarding human genetic modification technologies*, *Report*, Montreal: Genetics and Public Policy Centre.

Ledford, H. (2015) 'CRISPR the disruptor', *Nature*, 522(7554): 20–4.

Morrison, M. and de Saille, S. (2019) 'CRISPR in context: towards a socially responsible debate on embryo editing', *Palgrave Communications*, 5:110, https://doi.org/10.1057/s41599-019-0319-5

Savalescu, J. (2001) 'Procreative beneficence: why we should select the best children', *Bioethics*, 15(5–6): 413–26, https://onlinelibrary.wiley.com/doi/abs/10.1111/1467-8519.00251

United States President's Commission for the Study of Ethical Problems in Medicine and Biomedical and Behavioral Research (1982) *Splicing Life: A Report on the Social and Ethical Issues of Genetic Engineering with Human Beings*, Washington, DC: United States President's Commission.

SIXTEEN

Inclusive Education in the Post-COVID-19 World

Anna Pilson

Introduction

The COVID-19 pandemic has engendered and exposed inequality across society, and in few arenas more sharply than the education system. Disabled children have long been positioned as 'outliers' in a system created to embed neoliberal ideals – wherein success is presaged on a narrow concept of achievement, and anyone that cannot or does not meet these normative standards is deemed 'less than', and is, to adopt the parlance of critical disability studies, 'othered'. This is despite the fact that for the past two decades, inclusion has been considered a cornerstone of national educational policy in England.

Even prior to COVID-19, disabled children in England were found to be having a poorer educational experience than their peers. The House of Commons Education Committee's Inquiry into Special Educational Needs and Disability (SEND) in 2019 assessed the impact of the implementation of the Children and Families Act 2014, which was supposed to constitute the biggest reform of SEND education in a generation. Instead, the Inquiry identified not just structural issues (for example, inadequate funding, limited staffing capacity, absence of joined-up services) but also, in organizational and systemic behaviours failing to ensure accountability and inclusion, institutionalized ableism. The pandemic has served to exacerbate this trend, ushering in a raft of new practical, logistical and attitudinal barriers faced by disabled children. This active discrimination, perpetuated by the Coronavirus Act 2020, is symptomatic of long-term marginalization of disabled children in education, which focuses on making the child 'fit' normative systems, rather than being responsive to individual needs.

Therefore, despite its premise as a force for justice and equality, inclusive education often (unwittingly) perpetuates cycles of ableism: 'as a humanist orientation, inclusion privileges human traits (thought, capacity, sense-making)' (Naraian, 2020: 1). While this (at least superficially) egalitarian conception of humanism has fed into legislation that has sought to protect the rights of disabled children (for example, The UN Convention for the Rights of the Child, 1989; Children and Families Act 2014), it could also be argued that it fails to recognize the power and promise of individual difference, as well as our inevitable entanglement with non-human entities and systems. Therefore, this narrow concept of 'inclusion as fitting in' may, in actuality, be exclusionary.

This chapter expands on current and future conceptions of inclusive education, making suggestions as to what the COVID-19 pandemic has taught us that the education system might shed and embed in order to reframe inclusion. It asserts that if we embrace a posthuman orientation to our conception of inclusive education, then we move away from a rigidly stratified system, towards a fluidly intertwined relational, processual entity – an aspirational model for future society.

Inclusive education during COVID-19

It is useful to contextualize inclusive education in its current 'peri-COVID-19' manifestation, by examining how the response of the UK education sector to the pandemic impacted upon disabled children and their access to learning. Mirroring wider societal inequality, the education system in

England during the pandemic has been riven by injustice and unequal practice. Exemplifying this, in professed recognition of the difficult situation schools were facing, the government mandated a relaxation of Education Health Care Plan entitlements, which resulted in some local authorities claiming carte blanche to disregard the legally enshrined rights of many disabled children. Previously surreptitious institutional behaviour became openly (and legally) acceptable during the pandemic.

In the first national lockdown, annual reviews of progress were cancelled, assistive technology remained locked in school buildings, many children had no access to technology, adult support, differentiated resources, or resources provided in a suitable modified format (for example, Braille). The rapid move to the unknown entity of online learning platforms for the majority of educators made access arrangements the exception, not the rule. For example, many D/deaf children could not follow online lessons due to a lack of sign language interpretation. Others had to rely on automatically generated captions, the quality and clarity of which was questionable at best – giving rise to the nickname 'craptions' applied by some frustrated students. For pupils with mental health conditions, such as anxiety, the forced 'camera on' format of many lessons could be fatiguing and stressful.

Conversely, for some students, the move online provided a welcome respite from mandatory school attendance, with some expressing a preference for the often-asynchronous nature of the virtual learning experience. Online learning also allowed those with chronic illnesses, or those unable to attend school due to being deemed clinically extremely vulnerable, to keep up with the learning and social aspects of schooling. Online platforms have been utilized by the disabled community for many years, yet requests for their usage as reasonable adjustments have traditionally been met with refusal by gatekeepers until the pandemic forced the issue. Prior to COVID-19, inclusion was underpinned by a spatial precedent – physical presenteeism remained king.

Future conceptions of inclusive education

There continues to be no fixed universal definition of inclusion. While this could be dangerous, as it could lead to local authorities doing the 'bare minimum', it could also be freeing in post-COVID-19 education, as it may allow inclusion to become a more affirmative experience, and not a systemic and systematic 'tickbox' exercise that is 'done' to children. Educators and policymakers must position inclusion as a process – not an outcome. One that is malleable, reactive, proactive, osmotic. It should sheath each individual like a protective skin – responding to their needs – nurturing, not suffocating. In order to achieve this reframing, we need to consider what the education system needs to shed, and what it needs to embed, based on the lessons we are learning during the COVID-19 pandemic.

Firstly, it must embed accessibility – opportunities for asynchronous learning, and flexibility in terms of content and volume of curriculum. Being able to take control of learning (to an extent) has allowed many disabled young people to experience 'crip time' (Kafer, 2013: 27):

> Crip time is flex time not just expanded but exploded; it requires re-imagining our notions of what can and should happen in time, or recognizing how expectations of 'how long things take' are based on very particular minds and bodies. … Rather than bend disabled bodies and minds to meet the clock, crip time bends the clock to meet disabled bodies and minds.

The government's 'Progress 8' measurement of school performance in terms of ensuring pupil progress between Key Stages, has meant that temporality has become increasingly synonymous with achievement. This has meant that the curricular choices are more tightly controlled, leading to reduced flexibility to meet the holistic needs of some children. For example, visually impaired children often need to follow an additional curriculum, in which they learn independence skills, mobility skills, assistive technology skills. This historically would have required them to study fewer subjects in order for them to free up time within their timetable to undertake this learning. This is no longer permissible in many schools, who cannot afford for students to undertake fewer than 8 GCSEs in case it reflects negatively on the school's Progress 8 score. Such accountability measures directly undermine opportunities to access a more flexible and personalized curriculum.

Consequently, the education system must shed its obsession with normativity. Normative concepts of productivity and measures of attainment have proven themselves unnecessary. For example, the abandonment of formal external examinations for Key Stage 1, 2, 4 and 5 students in 2020 and 2021 has proven that terminal measures of attainment are of little 'real' value, and that schools and other institutions are capable of measuring progress in other (potentially more meaningful) ways. Additionally, normative constructs of space and optimal learning environments have been questioned by virtual learning. Many children have benefitted from a less rigid timetable, being able to go to the toilet without permission, to take a break from screens, to occasionally have time away from schooling during the day.

However, the education system in the UK, despite its outwardly professed commitment to equality, is inherently hierarchical. It is frightened of the autonomy virtual learning has provided for some children. This desire to return to classrooms, arguably before Science tells us it is safe to, is a control measure to preserve the power of education in creating a society that adheres to the Darwinian notion of 'survival of the fittest'. This exemplifies perfectly Braidotti's (2020: 466) claim that 'not all humans are equal and the human is not at all a neutral category'. Disabled children within the current education system have long been multiply marginalized by their age, disability, and achievement (or lack thereof) against standardized measures of success. Yet by forcibly removing education's adherence to previously inflexible conceptions of time and space, the pandemic may have given schools the opportunity to (re)model their practice via a posthuman, and thereby arguably more inclusive lens.

Conclusion

COVID-19 has undermined anthropocentrism. In destabilizing human supremacy, the virus has also undermined the dominance of neoliberal narratives of power and value, thereby giving educators licence to 'undo the current ways of doing – and then imagine, invent and do the doing differently' (Taylor, 2016: 6). The disruption of 'normalcy' has offered us an opportunity to reposition disability 'as an affirmative phenomenon: a chance to pause, re-jig and reorient education' (Goodley et al, 2019: 988), so that it is no longer in thrall to normativity.

By embracing educational practices during the pandemic that were previously perceived as the domain of the disabled, such as asynchronous, remote and virtual learning, we have automatically entered into posthuman assemblages – 'embedded, embodied and yet flowing in a web of relations with human and non-human others' (Braidotti, 2019: 4). Yet 'traditional' schooling, with its obedience to standardized outcomes and prescriptive methods, shies away from this more affirmative approach to teaching and learning. Policymakers must use the lessons taught to us by the crisis to reimagine inclusive education for post-COVID-19 times, via creating and centring pedagogies that embrace the intersectional identities of disabled children, rather than pigeonholing them as 'less than human' because their existence subverts entrenched normative expectations.

The existent narrative of exclusion and dehumanization, so insidious within the education system, can be inverted by adopting a posthuman approach to inclusive education going forward. In doing so, the disabled adults of tomorrow can (re)map the terrain of future society to recalibrate power dynamics – and to reimagine what 'being fully human' means. It may appear antithetical to adopt a posthuman approach when years of disability-centric legislation has attempted to position the disabled child on a path that would allow them to be recognized as human. But such acceptance is only achieved by attaining conventional markers of success (for example, passing exams, getting a job after leaving school). By embedding a hybrid model of learning, repositioning the human as a component of the educational assemblage, we can learn from disability's radical disruptiveness. Embracing posthumanism within the education system, therefore, may allow disabled young people to subvert 'taken-for-granted, ideological and normative under-girdings of what it means to be a valued citizen of society' (Goodley et al, 2014: 348) and navigate relational journeys in and across the 'new normal'.

References

Braidotti, R. (2019) 'A theoretical framework for the critical posthumanities', *Theory, Culture and Society,* 36(6): 31–61, http://journals.sagepub.com/doi/full/10.1177/0263276418771486

Braidotti, R. (2020) ' "We" are in this together, but we are not one and the same', *Bioethical Inquiry,* 17: 465–9, https://doi.org/10.1007/s11673-020-10017-8

Goodley, D., Lawthom, R. and Runswick Cole, K. (2014) 'Posthuman disability studies', *Subjectivity,* 7: 342–61, https://doi.org/10.1057/sub.2014.15

Goodley, D., Lawthom, R., Liddiard, K. and Runswick-Cole, K. (2019) 'Provocations for critical disability studies', *Disability and Society,* 34(6): 972–97, https://doi.org/10.1080/09687599.2019.1566889

Kafer, A. (2013) *Feminist, Queer, Crip*, Bloomington, IN: Indiana University Press.

Naraian, S. (2020) 'What can "inclusion" mean in the posthuman era?' *Journal of Disability Studies in Education*, 1: 1–21.

Taylor, C.A. (2016) 'Edu-crafting a cacophonous ecology: posthumanist research practices for education', in C.A. Taylor and C. Hughes (eds), *Posthuman Research Practices in Education*, Basingstoke: Palgrave MacMillan, pp 5–24.

SEVENTEEN

From TINA to TAMA: Social Futures and Democratic Dreaming in the Ruins of Capitalist Realism

Paul Graham Raven

The late Mark Fisher is perhaps best known for naming an experience or feeling which has been prevalent for the last decade or so, but is arguably older than that. What he labelled 'capitalist realism' (Fisher, 2009) can be summed up by Margaret Thatcher's famous claim that There Is No Alternative (TINA) to the econo-political regime which she, among others, ushered onto the world stage in the late 1970s. Another popular formulation is an aphorism variously attributed to Fredric Jameson and Slavoj Žižek, and popularized by Fisher's riffing on it: 'it is easier to imagine the end of the world than to imagine the end of capitalism'. Here I would argue that the end of the world – or at least the end of *a* world, of a particular way of life and its associated certainties – has become somewhat easier to imagine, given that we (though not everyone) have lived (or are perhaps still living) through just such an end.

I would not claim that the hegemony of TINA has been overturned as a result of the pandemic – though it was perhaps already weakening, as evidenced by the widespread retreat toward older (and often uglier) political programmes. Indeed, if we reformulate TINA in another way – as the bitter claim of 'no future' that was the rallying cry of the British punk explosion – then it is still very much with us, albeit announced not with the furious sneer of Johnny Rotten, but rather the exhausted resignation of a zero-hours employee whose earnings barely cover their rent and debts. Nonetheless, while it still feels like there is no alternative, it also feels like there *should* be, maybe even *must* be, an alternative. The spell has been weakened, if not yet broken.

We saw in the early months of the pandemic a period of mediated solidarity which crossed (but did not erase) lines of class, age, ethnicity and nationality; even the punditry – however implausibly – were busily swapping business-as-usual for everything's-changed-forever. However, that solidarity quickly fell apart. A stauncher Marxist than I might argue that COVID-19 provoked a sudden class consciousness: little was known beyond the fact that this disease was potentially lethal, regardless of who you were, where you lived, how much you earned. But it didn't last long, and debate will presumably rage for decades over the mechanism of reterritorialization. My money would be on the incredibly rapid way in which – typically of the capitalist-realist paradigm and its business ontology – the pandemic was turned into a competition between nations, between regions and cities, and eventually (inevitably) between individuals: a global biopolitical marketplace, complete with league tables of deaths-per-million, economic forecasts based on little more than wild guesswork, and the quantificatory simplification of hugely complex social and biological dynamics. And then there is the matter of a news media ecology optimized for clickworthiness *über alles* – a phenomenon at least as old as Baudrillard's controversial observation that the first Gulf war was a simulacrum (Baudrillard, 2009), if not older still, now so ubiquitous as to be effectively invisible. However, searching for causality in this mess is likely as futile as searching for causality regarding the coronavirus itself: no amount of data or argument will suffice to convince someone truly determined to nurture an alternative conclusion more congenial to their enduring episteme.

Speaking of epistemes, we should confront the discomforting truth that the momentary COVID-19 class was almost certainly not as universal as we might like to think – a chasm of affect all the more obvious in this later phase. Those who had something that passed for a 'normal' life before the pandemic are desperate to return to it (and I would be a liar if I claimed I didn't feel that pull myself); those for whom 'normality' was always a grim joke told about a luxury possessed by other people, meanwhile, seem to harbour an understandable resignation to whatever may happen, accompanied by an assumption that it will most likely be something that looks a lot like Business As Usual (which is TINA's twin brother). This is a survival mechanism, based on experience, and aimed at the avoidance of disappointment: these truly proletarian cohorts were probably not much caught up by the surge of solidarity, having seen similar come and go many times before, with little substantive change to show for it. (It is hard to be certain, of course, as few if any media outlets would have bothered to ask them – and any answer in the negative would have spoiled the moment.) Furthermore, the inequities that existed before COVID-19 have not gone away. If anything, many have been reinforced by it – and it is mostly those who felt the solidarity most strongly who are the likely beneficiaries of that reinforcement.

All of which may sound a little hopeless – and at time of writing (July 2021), it would be forgivable to have lost hope that the pandemic might change anything fundamental about the way we live, collectively speaking. But nonetheless, there was a moment in which the Global-Northern middle classes were at least saying that things needed to change, must change, were changing … and that's not nothing. That it took a presumed-to-be-universal alien threat to achieve that consciousness is notable, not least because it implicitly endorses the more-than-a-little-fascist thesis of Adrian 'Ozymandias' Veidt in Alan Moore and Dave Gibbon's original *Watchmen* comics: the easiest way to unify a collection of humans is through the construction of a non-human (or simply dehumanized) Other to serve as an enemy. (All those martial metaphors in pandemic news coverage are no accident, even if they were not necessarily deliberate.) Therefore, as wiser commentators than I have already observed, anyone hoping that the pandemic might serve as some sort of dress rehearsal for the climate crisis – of which COVID-19 might best be thought as a harbinger – would be wise to reassess their assumptions. Capital, the very engine of the global factory from which these threats issue forth, knows all too well how to divide us, and thus to conquer us.

Here I am expected, if not exactly obliged, to offer a solution for these problems. (The expectation that critique must come with a solution is another feature of capitalist realism, according to which no document is permitted to end without a sales pitch.) I have no solution to offer – and frankly the pandemic has made me even more distrustful of those who peddle 'solutions' than I was before. Nonetheless, to restate the point: the star of hope in the Pandora's box of the pandemic was the demonstration that the imprisoning narrative of capitalist-realist futurity can be escaped. But we have also seen that the threat of an abstracted Other will not suffice to hold together any class consciousness thus engendered: the momentary COVID-19 class was a top-down narrative, and those can always be recuperated by those for whom top-down narratives are their stock in trade. We cannot win on that battlefield, though we might achieve temporary deterritorializations in some corners of it.

So, no solutions. I do have *suggestions*, however, though I can offer them only as a form of faith – and one that stands very much counter to the technocratic metanarrative of capitalist realism – rather than a promise. With that disclaimer made, I believe that the formation of a new metaclass, subjectivized by climate change or pandemics or poverty – all of which are externalities of capitalism's increasingly amok trajectory – must be built from the ground up rather than from the top down, built by reference not to these abstract problematics, but rather to their concrete, situated expression in the lives of communities. And despite my diagnosis, which may well read as hopeless, it is precisely through the nurturing of hope that I propose to achieve it.

That is to differentiate clearly between hope and mere optimism – a distinction of some standing in the field of utopian studies, most recently articulated by Phillip Wegner (2020). With apologies for the reductiveness of this summary: optimism is passive, taking the form of a Whig-historical assumption that things *will* get better, which in turn absolves the optimist from having to do anything other than stay optimistic; hope, meanwhile, is active, taking the form of a conviction that things *could* get better, if only enough people are willing to work to make it so. Under the hegemony of TINA, even imagining that things might be substantively different to the continuity future on offer is effectively forbidden, policed through mechanisms of humiliation, ridicule and scorn. As such, Wegner argues, dreaming itself becomes a utopian act of defiance: utopia is not just the better future you imagine, but also the very act of imagining it.

My own work, and that of many colleagues, might be thought of as a utopian praxis: believing that things could be better – *should* be better, *must* be better – in times to come, while acknowledging that they can never be perfect, and deciding to work toward better anyway. This aim of exploring possibilities, of opening up the complexity of sociotechnical change to discussion by non-experts, is achieved through imaginative and creative work, informed by the best scientific knowledge we have available. It is the work of science fiction, in other words – but done not so much for the sake of entertaining its audience as for inviting them into futurity as a space where their own ideas, experiences and values might be expressed and taken into account. (That said, we find that making it entertaining helps a lot.)

While we distinguish our futuring from the hegemonic forms in terms of our epistemological assumptions, we are also working to distinguish it in terms of its methodology, and its constituency. In concrete terms, this means methods and practices such as co-production, action research, social practice placemaking, transitions design, narrative prototyping, design fiction, experiential futures … *social* futures, developed not just *for* the social but *with* the social, *by* the social. The aim is the concretization, situating, and democratization of futuring (Raven, 2021): the production of futures in which ordinary people might see themselves and their concerns, their fears and their dreams, because they have been asked to put them there.

Because There Are Many Alternatives, and always have been – far more than you or I or any one person might be able to imagine. Our futures will perforce be lived together; as such they should – they *must* – be dreamed, debated, remixed, (re)made together, too. The pandemic has weakened the spell of capitalist realism, but the questions to be asked about those futures are not only big abstract questions of epidemiology and vaccines and lockdowns, or of climate change and inequality. We also need to ask the small questions, which are just as important: the good life is as much about meals with family, coffees with colleagues, and beers with friends, as it is about political principles and economic programmes. Social futuring – at least as I hope to practice it – sees the former, smaller questions as the route to answering the latter, larger ones.

The time is right, and the theories and methodologies for this collective effort are ready to hand. Likewise the ideas and experience and expertise of ordinary people – experts in their own lives and locales – waits for us to combine it with our own, not as academics or elites or 'thought leaders' or 'disruptors', but as fellow passengers aboard a planetary ship sailing an ocean of time.

Any future is always already a story, but there are as many narratives as there are narrators, and the established style of storytelling has brought us close to disaster. It's time to broaden the circle, and thus to return the question 'how might we live?' to its true constituency.

References

Baudrillard, J. (2009) [1991] *The Gulf War Did Not Take Place*, Sydney: Power Publications.

Fisher, M. (2009) *Capitalist Realism: Is There No Alternative?* Ropley: Zero Books.

Raven, P.G. (2021) 'Concretise, situate, democratise: the museum of carbon ruins', *Transforming Society*, 16 February, http://www.transformingsociety.co.uk/2021/02/16/concretise-situate-democratise-the-museum-of-carbon-ruins/

Wegner, P.E. (2020) *Invoking Hope: Theory and Utopia in Dark Times*, Minnesota, MN: University of Minnesota.

Conclusion: Thinking about 'the Human' during COVID-19 Times

Paul Martin, Stevienna de Saille, Kirsty Liddiard
and Warren Pearce

In this final chapter we draw together some of the main themes emerging from the various chapters and reflect on what this tells us about being human in COVID-19 times. As outlined in the introduction, these essays have focused on three key issues during the pandemic that are fundamentally concerned with the experience, meaning and understanding of being human. Firstly, the marginalization of many groups of people and how they are de/valued in the response to the virus. Secondly, the role of new scientific knowledge and other forms of expertise in these processes of inclusion and exclusion. Thirdly, the remaking and reordering of society as a result of the pandemic and the opening up of new futures for work, the environment, culture and daily life. These themes were considered in the four sections of the collection, and the main points from each are summarized here, before a final consideration is offered on what this tells us about being human during and after the pandemic.

Knowing humans

This collection of essays starts by exploring how COVID-19 has been known and represented in different metaphors, models, representations, and media, as the pandemic has unfolded. In analysing these processes, new insights are provided about how we understand the human. While the virus was the same molecular structure the world over (at least before the onset of variants), this section shows the myriad of different methods and resources by which the resulting disease and its impacts became known to policymakers, professionals and publics, and how these differed across the world. Three key features of this emerge. Firstly, whether through science, metaphor or imagery, the ways in which COVID-19 became known could both exacerbate existing inequalities or provide the means to counter them (Nerlich; Ballo and Pearce; Rosvik et al). In this sense, they form the ground for contestation over the meaning of COVID-19. Secondly, citizens found themselves dislocated from established sources of knowledge about the virus, which they felt to be either incomplete or inadequate (Garcia; Vicari and Yang; Røstvik et al). These uncertainties about what risks they faced, how to respond, and their responsibilities to self and others, fed into high levels of distrust and confusion. Thirdly, social media platforms provided important avenues for the dissemination and contestation of COVID-19 knowledge which, while theoretically global in nature, is embedded within the values of companies that exist in specific political and cultural locales (Vicari and Yang). The interaction between different lay and expert representations of COVID-19, the uncertainties citizens faced, and the central role of social media in dissemination and controversy, have been a defining characteristic of the pandemic. These have helped underpin both the shared human experience felt across the globe and the major differences in responses and consequences in different countries. The result has been a powerful set of discourses about our common humanity, and solidarity between different social groups and geographical locations. At the same time, however, the dynamics of metaphor, model, image and media traced in this section have also led to fragmentation and othering, allowing authoritarian and nationalist leaders to play the blame game. This tension between shared experience and fragmentation and othering has only intensified as the pandemic has progressed.

Marginalized humans

The next section explored how COVID-19 has exposed the ways in which key marginalized groups – Black, poor, displaced and/or disabled – have been rendered disposable as a result of systemic racism, toxic capitalism, and austere underfunding of health, education and social care. This has created new vulnerabilities for marginalized people, while governmental, science and policy responses have failed to properly protect them. It has had several significant consequences, with COVID-19 exacerbating and intensifying existing inequalities of access and precarities in work, health and education. As these chapters demonstrate, these processes of marginalization have been especially hard felt by disabled people and their families, as the global pandemic has heightened the worst effects of ableism and disablism in their lives (Liddiard et al; Goodley and Runswick-Cole). The intersectional and transdisciplinary approach adopted here illuminates the diverse ways in which marginalized groups, particularly those labelled disabled, have been further devalued and dehumanized with the pandemic response. These include deeply embedded cultural assumptions about the value of different human lives (Liddiard et al; Goodley and Runswick-Cole; Titchkosky), and policies and practices that systematically exclude individuals and groups on the basis of identity or ability (Pandey and Sharma; Liddiard et al). In response, the contributors call for theory and practice that address intersectional forms of dehumanization at all levels (Liddiard et al; Goodley and Runswick-Cole; Titchkosky) as pandemic emotions of shame, exclusion, othering, fear, a lack of protection, dehumanization, degradation, and disposability, are being felt by many. An important dimension of this approach is to incorporate studies of affect into our analyses of the impacts of COVID-19 upon marginalized people. To achieve this, there is a need to understand affect in ways that are sensitive to the complex interplay between social, cultural, historical and economic conditions (Pandey and Sharma; Liddiard et al; Goodley and Runswick-Cole;

Titchkosky). Only by doing this can the dynamics of exclusion and inequality, and the experiences of marginalized humans, be fully understood and acted on.

Biosocial humans

The role of bioscientific knowledge, and the practices based on its claims, have guided the management of health services, the experience of patients and citizens, and framed our understanding of the human condition during COVID-19 times. The pandemic has been marked by the daily circulation of new knowledge about the virus, and our biological and social response to it, both at individual and population level. We have learnt much about viral replication, incubation, transmission, mutation and variation, with this being shared widely among both experts and lay publics. We might even think of this in terms of a 'popular virology'. New discourses of susceptibility and resistance have mingled alongside understandings of how we can best protect ourselves. As part of this, vital (and vitalist) decisions have been made about who should be protected, who should be treated and who might be left to die. Ways of measuring human vulnerability, and policies about the allocation of scarce resources have been driven by a utilitarian logic and combined into biomedical practices that marginalize and discriminate against the elderly and the disabled (Goodwin et al). These reflect wider cultural beliefs that devalue the lives of ageing humans; a social fact that is manifest in the lack of care taken by governments in protecting vulnerable people in many care settings (Michalko). In this context the shift to online channels of communication has made the fight against discrimination within biomedicine all the harder, and privileged some ways of being in – and knowing – the world over others. This is well illustrated in the experience of blind and Black people as they struggle to stay connected to their communities (Michalko, Williams). An important factor shaping both these experiences and our changing knowledge of COVID-19 is the development and diffusion of new technology, most notably vaccines. Scholars working in STS see technoscientific change as not simply the unfolding of some inner technical logic but rather the result of complex interactions between science, culture, economics and politics. This is clearly visible in the development and application of gene sequencing infrastructures that have enabled the tracking of the virus and its variants. These technologies are being developed in ways that increasingly frame our understanding of human health as biologically rather than socially determined, offering promises of biomedical and pharmaceutical advances to treat our future health needs as if independent of social location (Martin).

Human futures

The final section of the collection extends the preceding themes. Rather than trying to predict what might come next in a post-COVID-19 world, the authors explore how specific futures are being made in the here and now, and how they open up or close down the possibility of a more equitable, sustainable and inclusive society. Such futuring can take many forms. As highlighted, these can be technoscientific and involve the creation of new imaginaries that might serve a performative function in guiding the development and deployment of emerging technologies. These are sites of contestation where different visions of human futures are forged as part of struggles over how to manage the pandemic. In some futures technical fixes are proposed to protect a limited number of people from infection (for example, via gene editing future generations) rather than adopting universal public health measures to protect everyone (Morrison). Such struggles reflect the uneven distribution of power and resources already clearly visible in the global distribution of vaccines.

In the early stage of the pandemic there appeared to be an opening up of possibilities, with new social futures that were more equitable, sustainable and inclusive seeming possible, but as the old order has started to reassert itself these have quickly reverted to futures based on competition between individuals and nation–states (Raven). The case study of inclusive education in the UK (Pilson) highlights a series of contradictions and tensions as COVID-19 has opened new ways of teaching. These hold the possibility of really improving the lives of children with learning disabilities, a group often seen as less than fully human. However, as some sort of normality returns, the use of online platforms has also enabled new forms of exclusion. In this context the politics of hope and alterity are vital. As Raven points out, hope is not just wishful thinking but can be a programmatic response to crisis that creates new human possibilities. The end of business as usual, and the dramatic response from many governments to the onset of the pandemic, clearly showed that another world is possible. These might include novel ways of organizing social and economic life that are not based on market-driven destruction of the environment and depleting the earth's resources, but instead promote planetary wellbeing and human flourishing. Ideas such as Responsible Stagnation (Arthur et al) help connect the many local social innovations that have sprung up across the globe during the pandemic which seek just such a future. In imagining a better world, we also reimagine what it might mean to be human after the pandemic. This collection has brought together a range of different approaches and perspectives on both COVID-19 and the human. Some are personal, while others are based on current research. A number of chapters are conceptual, while others are descriptive. This diversity of topics, styles and approaches is deliberate, as we want to reflect the richness and variety of conceptions of how we might understand the human. We consider this a strength, not a weakness, as we reject a single dominant idea of human identity and experience. As mentioned in the introduction, many of the contributors are critical of the narrow Western concept of the human and seek to embrace a much broader understanding, based on a diversity of experiences and standpoints.

Within this framework the collection has also reflected the two main disciplinary perspectives within the iHuman project: critical disability studies (CDS) and science and technology studies (STS). The focus on disability strives to redress the neglect of this topic and the voices of disabled people within mainstream debates on the pandemic. In doing so, it goes much further than simply describing the impact of marginalization and disposability, and instead provides a powerful analytical lens that takes disability as a starting point from which to rethink many aspects of so-called normal life. In this sense, disability becomes a driving subject of inquiry and a means of unsettling established ways of seeing the world and framing problems and their solutions.

In a similar vein, STS places an emphasis on the production of knowledge and the role of science and technology in the constitution of contemporary society. While this is increasingly reflected within mainstream thinking, it also challenges the idea that technologies are neutral and that innovation is always desirable and inevitable, and critically explores the relationship between knowledge and power. In both CDS and STS there is an emphasis on the importance of our entanglement with non-humans and the politics of alterity – the idea that things *could* be different, that we *can* build human futures that are more inclusive, equitable and sustainable *if we choose to do so*. COVID-19 has demonstrated that alternatives are indeed possible.

What then has this collection of essays told us about being human during and, possibly, after COVID-19?

The authors have outlined the way in which ideas of the human are being constructed in different social, cultural, scientific, and technological domains during the pandemic by new knowledge, government action, and cultural norms. The pandemic has helped cement a view of the human as collective, prosocial, and sharing a common bond between all people. This shared experience of living (and dying) during COVID-19 has proved a unifying force. At the same time, biological, social and economic differences have been made more visible, in particular, differential access to essential resources such as safe housing, a stable income, and vaccines, lack of which has increased the marginalization of Black, poor, old and disabled people in many (if not all) countries, regardless of national wealth.

Many of the essays have drawn on a critical posthumanities ethos to highlight both the dominance of Western-centric ideas of the human, and alternative notions of human identity in different social settings, cultures and parts of the world. Such differences are visible in ideas about collective versus individual responsibility, about who or what is the cause of the pandemic, and how we might best protect ourselves. Such understandings are powerfully shaped by media, scientific claims and political ideologies, many of which are anchored in national institutions and cultures.

So, is it possible to talk about the post-COVID-19 human, or even consider what sort of human we want to become? In trying to answer this, it must be noted that this collection of essays is motivated in part by a desire to make visible the possibilities for a more just, inclusive and sustainable world, and therefore emphasizes the final section's central message that there are alternatives. It is possible, even urgent and essential, to reconstitute and reimagine what it means to be human in a way that emphasizes our common humanity, our collective nature and our sociable character. This is in opposition to dominant neoliberal or authoritarian popularist narratives that stress division and competition. In these futures there is no alternative to the existing social, economic, and ecological order.

This is where the practical politics of hope becomes important. As we move from the peri- to the post-COVID-19 moment, there is still an opportunity to consolidate some of the beneficial possibilities opened up by the pandemic. We have seen governments, businesses, and communities across the world act in positive, supportive ways that were unimaginable in February 2020. This will remain in the collective memory for some time to provide a potential buttress against a return to business as usual. The figure of this equity-seeking, sociable, *generous* post-COVID-19 human is one we need to promote through a politics of hope and liberation. As other crises loom large and the climate emergency deepens, this is surely the most important lesson we can learn from COVID-19 times.

Index

Note: References to figures appear in *italic* type;
those in **bold** type refer to tables. References
to endnotes show both the page number and the note number (43n2).